T0338256

The Basics of Troubleshooting in Plastics Processing

Scrivener Publishing
3 Winter Street, Suite 3
Salem, MA 01970

Scrivener Publishing Collections Editors

James E. R. Couper	Ken Dragoon
Richard Erdlac	Rafiq Islam
Pradip Khaladkar	Vitthal Kulkarni
Norman Lieberman	Peter Martin
W. Kent Muhlbauer	Andrew Y. C. Nee
S. A. Sherif	James G. Speight

Publishers at Scrivener
Martin Scrivener (martin@scrivenerpublishing.com)
Phillip Carmical (pcarmical@scrivenerpublishing.com)

The Basics of Troubleshooting in Plastics Processing

An Introductory Practical Guide

Muralisrinivasan Natamai
Subramanian

Scrivener

Co-published by John Wiley & Sons, Inc. Hoboken, New Jersey, and Scrivener Publishing LLC, Salem, Massachusetts.
Published simultaneously in Canada.

For general information on our other products and services or for technical support, please contact our Customer Care Department within the United States at (800) 762-2974, outside the United States at (317) 572-3993 or fax (317) 572-4002.

Wiley also publishes its books in a variety of electronic formats. Some content that appears in print may not be available in electronic formats. For more information about Wiley products, visit our web site at www.wiley.com.

For more information about Scrivener products please visit www.scrivenerpublishing.com.

Cover design by Russell Richardson.

Library of Congress Cataloging-in-Publication Data:

ISBN 978-0-470-62606-1

Printed in the United States of America

10 9 8 7 6 5 4 3 2 1

Contents

Preface

Plastics processing is a core technology in major sectors of the plastics industry. In spite of the growing importance of the field of plastics processing, a clear and uniform practical guide covering the entire field of troubleshooting of plastics processing has not been presented until now.

The Basics of Troubleshooting in Plastics Processing: An Introductory Practical Guide will appeal to all those who are involved in the plastics production sector. The material in the book covers both fundamental and practical aspects of plastics processing and attempts to provide the necessary background to understand the factors that constitute successful plastics manufacturing.

In an effort to mirror the goals of the plastics industry, the scope of the book emphasizes the importance of the high quality production of end products, processing, while deliberately restricting coverage of machine details to the main processing technologies. These technologies are: injection molding, extrusion, extrusion blow molding, injection blow molding, thermoforming, and rotational molding. By including fairly comprehensive details of technical information and reference material, this ensures the book is suitable for classroom and industry training purposes, without affecting its overall usefulness as a technical and reference work. It will be very suitable for research workers, engineers and operators in industry, students in plastics processing, as well as to all those seeking an introduction in plastics processing.

The framework of the book underscores both the book's academic and professional aspirations. Thermoplastics materials and characteristics are discussed in Chapter 2. A useful introduction of additives is covered in Chapter 3. The plastics processing techniques occupy Chapter 4 which also gives information pertinent to troubleshooting. Chapter 5 deals specifically with troubleshooting problems and solutions and elucidates the various control measures available in plastics processing. Chapter 6 briefly presents the future trends related to plastics.

My major objective in writing this book was to provide a thorough background in plastics processing which is particularly important in order to increase the productivity and reduce wastage. Additionally, I hope this book will help people to develop the skills necessary to solve the problems in a stepwise manner.

I would like to thank Dr. A. Thamaraichelvan, and Mr. A.N. Sapthagireesan for their encouragement. In addition, special thanks are due to my wife and sons for their support.

I am especially grateful to Mr. Martin Scrivener and editorial staff at Scrivener Publishing for an excellent and professional job.

Muralisrinivasan Natamai Subramanian

November 2010

1

Introduction

Plastics have become an integral part of our lives. Plastics are an excellent and very useful material and they are functional, hygienic, light, and economical. Using a key polymer processing system, plastics produce diverse products used in packaging, automotive and industrial applications, and also extensively used in medical delivery systems, artificial implants and other healthcare applications, water desalination, and removal of bacteria, etc.

Until the 1930s and in early 1940s, thermoplastics were not common material. Ever since the first industrial scale production of plastics (synthetic polymers) took place in the 1940s, the production and consumption has increased considerably. Although plastic materials are relatively new, they have become basic and indispensable in our life with different shapes, sizes, and applications that can be seen daily at home, office, and even on the street.

The growth in the manufacture of thermoplastic products for various applications has been considerably increased.

It includes many light engineering applications. The plastics materials usage for food packaging has obvious advantages associated with the non-toxic nature of these materials and their resistance to chemical and biological degradation [1].

1.1.   Market Trends

Today, the requirements are constantly changing and rising to higher levels [2]. The world's annual consumption of polymer materials has increased from around 5 million metric tons in the 1950s to nearly 100 million metric tons today [3, 4]. The worldwide demand for plastic pipes is forecast to increase 4.6% annually through 2012 to 8.2 billion meters or 18.2 million tons. Demand for HDPE (polyethylene) pipes will benefit from use as small-diameter pipes in natural gas transmission, as conduit for electrical and telecommunications applications, and as corrugated pipes for drains and sewers [5].

The US market for plastic healthcare packaging is expected to reach 3.8 billion pounds of products in 2010. This amount is forecast to increase to nearly 5 billion pounds in 2015. PP (polypropylene) packaging, the largest segment of the market, will reach 1.3 billion pounds in 2015, after increasing by 6.3% per annum from the estimated 2010 total of nearly 1 billion pounds [6].

Global demand for PE, the largest-volume basic polymer, is expected to grow about 4.2% per year to reach about 81 million metric tons by 2013. Demand for engineering plastics will rise by 3.1% per year to reach 5.4 billion lbs by 2012. Polycarbonate, nylon, and ABS will continue to be the largest-volume engineering plastics, accounting for more than 75% of total demand by 2012 [7].

Polymer demand has been driven by high levels of investment particularly in packaging, appliances, consumer electronics, and automotive [8]. However, in the modern global market, quality is a key issue to remain competitive in business. Quality can no longer be simply the result of an inspection process, but very much part of the strategic planning process of successful companies [9].

1.2   Importance of Plastics

Plastics are increasingly important in the manufacture of materials due to their significant higher strength to weight ratio and stiffness, as well as impact strength. The main drivers for the rapid increase in their use are low cost and the possibility of achieving total recyclability. But the large scale and widespread use of plastics is due to its low density and ease of processing.

Plastics are macromolecules derived from monomers, also called polymer. The word "poly" meaning many and "mer" designating the nature of the repeat unit [10, 11]. Polymers are from synthetic or naturally occurring material which can be used with modification to suit with respect to processing. The term "polymer" is these days known as "plastics" when referring to macromolecules like polypropylene, polystyrene, etc.

Plastics are constructed by the covalent linking of simple molecular repeated units [12]. Plastics are composed of carbon, nitrogen, oxygen, sulfur, chlorine, fluorine, and silicon. Moreover, plastics are made from petrochemical products which are a rich source of methane, ethylene, aliphatic, and aromatics. Variations in the elements make the plastic into stiff or flexible, linear or branched, and hard or soft.

Plastics are classified based on recyclability into thermoplastics and thermosets. Polyethylene (PE), polypropylene (PP), polyvinylchloride (PVC), etc., are some of the thermoplastics and phenol formaldehyde, urea formaldehyde, etc., are examples of the thermosets. Both thermoplastics and thermosetting materials may be molded and then cooled to obtain the end product. Thermoset once molded cannot be either softened or reprocessed.

Thermosets lead to products which are not recyclable. Moreover, it will form a network and it can neither be melted nor reprocessed. Once shaped, it can be altered by post forming operations if required. Pre-polymers are to be made before processing in thermoset processing. However, thermoplastics soften while heating and solidify during the cooling process. Thermoplastics can be recycled by either direct heating or after grinding into granules of scrap products [13].

Processing technology that shapes material and technology of plastics allows the manufacture of parts with lightweight, precision and strength, and low cost. It is cheaper than metal or ceramic processing. However, to use plastic effectively and to have the best advantage of its application, specific characteristics or physical properties must be considered.

In plastics processing, with technology and application advances, conventional product replacement and unlimited innovation can take place. Plastic raw materials are also widening its range of products. Achieving higher performance with increased quality is the major challenge in manufacturing today. Plastics processing, therefore, requires constant and sometimes fundamental change.

Moreover, as plastics have replaced many conventional materials, such as metal and wood, in many applications throughout the world, the growth will be accelerated by the tendency to substitute plastics for metal [14].

1.2.1   Plastics vs Metal

When plastics are compared with metals, some of the properties of plastics can be considered either favorable or unfavorable depending upon the application. Plastics are not so strong as metal. However, plastics have certain properties to be considered as advantageous for engineering applications. Plastics have better chemical and moisture resistance. Plastics are more resistant to shock and vibration than metals. Plastics are usually easier to fabricate than metals. Nylon material is self-lubricating and does not require any external lubrication during operation.

Table 1.1 Materials properties comparison.

No.	Property	Plastics	Metals	Paper	Wood
1.	Density	Low	High	Low	Low
2.	Mechanical properties	Better	Good	Poor	Poor
3.	Chemical Properties	Good	Better	Poor	Poor
4.	Water resistance	Good	Corrosive	Absorb	Absorb
5.	Shock and vibration resistance	Good	Better	Poor	Poor
6.	Microbial resistance	Good	Poor	Poor	Poor
7.	Degradation	Difficult	Easy	Easy	Easy

1.2.2 Plastics vs Paper and Paper Board

Paper and paperboard are widely used as food packaging materials and have been used with a number of chemicals such as slimicides, bleaching agents, and inks during the production process. Virgin paper and paperboard products produced by pulping, bleaching, and treatment processes undergo severe chemical treatment and it is impossible to eliminate the chemical residue present [15–17]. Hence, migration of chemicals from paper packaging to the food is quick resulting in toxicity to humans as being the main concern.

1.3 Plastics Processing

Plastics processing requires the knowledge fundamentals of the raw material, additives, process control, and finally the product properties required to the finished end product. Today, polymer contains a package of ingredients to modify its properties while processing, or at its end product stage to create a new one.

In thermoplastics, processing techniques can be classified into either batch or continuous process. Batch processes include injection molding, thermoforming and rotomolding. Extrusion of plastics is a continuous process. However, blow molding is available both in batch and continuous process. In these days, online continuous thermoforming machines are available along with extrusion process.

As the scientific techniques become available, the plastics processing is quickly incorporating the changes. However, new solutions pose new problems so these continue to be challenges to overcome. Troubleshooting helps to solve the problem at the root and increase the production efficiency during processing.

1.4   Fundamentals

- Based on recyclability, plastics can be divided into thermoplastics and thermosets.
- Knowledge of properties with respect to plastics raw material or its end product is essential to establish the trouble free plastics processing.

References

1. Plastics versus Food Contamination, *Corrosion Technology*, June 1965.
2. Reilly, J.F., Doyle, M., Kazmer, D., An assessment of dynamic feed control in modular tooling, *J. Inject. Mold. Technol.* 5 (1) (2001) 49–59.
3. Takoungsakdakun, T., Pongstabodee, S., 2007. Separation of mixed post-consumer PET–POM–PVC plastic waste using selective flotation. *Separation Purification Technology* 54 (2), 248–252.
4. Burat, F. Güney, A. and Olgaç Kangal, M../ *Waste Management* 29 (2009) 1807–1813.
5. *Macplas International*, March 2009, p. 20.
6. *Plastics and Rubber Weekly*, 28th May 2010, p. 11.
7. *Chemical Week*, 171, No.16, 8th-15th June 2009, p. 22–26.
8. *International Bottler and Packer*, 83, No.5, May 2009, p. 12–13.
9. McKeown, P., (1992), "Implementing quality improvement programmes", Robotics & *Computer Integrated Manufacturing*, Vol. 9 No. 4/5, pp. 311–20.
10. Mark, H. and Whitby, G.S., (eds), *Collected Papers of Wallace Hume Carothers on High Polymeric Substances*, John Wiley & Sons, New York, 1940.

11. P.J. Flory, *Principles of Polymer Chemistry*, Cornell University Press, Ithaca, NY, 1953.
12. H. Staudinger, *From Organic Chemistry to Macromolecules*, John Wiley & Sons, New York, 1970; H. Staudinger, Chem. Ber., 1924, 57, 1203.
13. Throne J.L. *Adv Polym Technol* 1987;7(4):347.
14. Nuñez, A.J., Sturm, P.C., Kenny, J.M., Aranguren, M.I., Marcovich, N.E., Reboredo, M.M. Mechanical characterization of polypropylene-wood flour composites. *J Appl Polym Sci* 2003;88(6):1420–8.
15. Ozaki, A.,Yamaguchi, Y., Fujita, T., Kuroda, K., Endo, G., Chemical analysis and genotoxicological safety assessment of paper and paperboard used for food packaging, *Food and Chemical Toxicology* 42 (2004) 1323–1337.
16. Arvanitoyannis, I.S., Bosnea, L., Migration of substances from food packaging materials to foods, Crit. *Rev. Food Sci. Nutr.* 44 (2004) 63–76.
17. Vitrac, O. Mougharbel, A., Feigenbaum, A., Interfacial mass transport properties which control the migration of packaging constituents into foodstuffs, *J. Food Eng.* 79 (2007) 1048–1064.

2

Plastics Materials

Plastics material undergoes different and complicated thermo-mechanical processes. It experiences significant change in rheological, mechanical, and transport properties due to large variations and rapid cooling. The term "polymer" is preferred to "resin" when referring to a high molecular weight substance like polystyrene or polypropylene.

Plastics are:

- Made from chemical raw material composed of atoms of carbon in combination with other elements [1, 2] called monomers, which are basic materials including those made from coal, alcohol, natural gas and petroleum.
- Made up by the repeated addition of one or more types of monomeric units.

Plastics have progressed with invention efficiently and products can be manufactured economically. But various plastics processing and its equipment differ substantially. It can range from cups for drink dispensers to key components in the aerospace industry, from toothbrushes to telephones, from computers to cars, from electrical appliances to automotive parts, from ballpoint pens to word-processor housing [3–6]. Plastics are also used in a wide range of applications such as clothing, housing materials, medical application, etc.

Plastic materials have become basic and indispensable in our life. To protect against contamination and conserve them, food products are distributed in different plastic packages: bags, bottles, boxes, etc. that contain all kinds of edible products: liquid (water, milk, cold beverages) or solid (fruit, meat, fish, frozen foods, etc.). The group of commercial plastics, also termed commodity plastics, consists of the most used polymers in terms of volume and number of applications. They are mainly polystyrene (PS), polypropylene (PP), high- and low-density polyethylene (HDPE, LDPE), polyethylene terephthalate (PET) and, in lower proportion, polycarbonate (PC) [7].

Plastics are available in the form of pellets, granules, and powder which can be extruded, blow molded, injection molded or rotomolded to fabricate products. In the global markets over many years, plastics have proven their utility and cost effectiveness. Even though the markets are mature, applications are continually being developed. New applications and developing markets are expected to increase the growth of plastics [8].

Plastics can offer impressive advantages [9]. They are light in weight, with good strength to weight ratio, cost effective, and corrosion resistant. However, plastics are not very cheap although plastic products can be cheaper than metal products. They can be shaped and mass produced with design freedom and reduce assembly time with ease and speed. They are good electrical insulators and colorful. Today, plastics are widely used in many important emerging technologies.

Modern polyethylene resins and several other thermoplastic materials including poly(vinylchloride), acrylic polymers are characterized by a narrow molecular weight distribution and advantageous mechanical properties, but are subjected to surface deterioration or "melt-fracture" during processing by extrusion [10], which contains the understanding and control of the "melt-fracture" phenomenon.

2.1   Properties and Processing

Today, a great deal of interest is placed on the use of plastics because of lower cost, and because they are a major source for industrial products and abundantly available. Plastic also has electric resistance, lower thermal conductivity, ability to come in an unlimited range of colors, and excellent surface finish [11].

Physical properties of the polymer are primarily dependent on the crystalline, amorphous or molecular orientation of material. This morphology is in terms dependent on thermal and mechanical treatment during processing [12]. The materials physical properties that help in processing are melt temperature (Tm), degradation temperature (Td), glass transition temperature (Tg), processing temperature, and density.

Plastics materials undergo different and complicated thermo-mechanical processes and experience the significant change in their rheological, mechanical, and transport properties due to large pressure variations and rapid cooling. Semi-crystalline plastics undergo a phase transition during heating or cooling near the melting point as demonstrated for PS and polyacetal. The reason for the behavior of semi-crystalline material is due to higher and necessary processing temperature during injection molding [13]. Amorphous materials generally provide better weld line strength than semi-crystalline materials, and a polymer material with higher flow rate may allow better packing for a stronger weld line.

Polymer properties are strongly influenced by [14]:

- Molecular weight (MW) of the material
- Molecular weight distribution (MWD)
- Degree of branching

2.1.1   Molecular Weight

The effect of increasing MW is the opposite of broadening MWD as when polymers vary in both MW and MWD [15]. Higher molecular weight plastics give stronger physical properties. However, they can be harder to process. Increase in molecular weight increases important parameters such as impact resistance, wear resistance, high deflection temperature, and stress crack resistance. The physical and mechanical properties are affected by the molecular weight distribution (MWD).

2.1.2   Molecular Weight Distribution (MWD)

With change in MWD, processing related problems can occur and fundamental improvement in processing conditions may need to have a product. The shape of the molecular weight distribution and the average molecular weight are directly related to the melt flow properties and processing performance, in terms of extrusion output rate and pressure [16]. The length of the molecule increases as the resistance to flow increases. As the breadth of the distribution increases there is a greater tendency to shear thinning [17].

Variations in molecular weight distribution would be expected to affect the temperature dependence of melt elasticity. Problems associated with polymer processing can occur if there is a change in the molecular weight distribution of the raw material.

2.1.3   Flow Properties

Flow properties are measures of viscosity and visco-elasticity of the molten polymer. There are two primary flow mechanisms, shear and elongational flow. Shear flow describes the response of the polymer to an imposed shear forcing. Elongational flow is an imposed stretching or pulling force. Although not independent of each other, they both have their own viscosity measures [18].

2.1.4   Degree of Crystallinity

Material with a higher degree of crystalline nature tends to be brittle because of the weak crystal-crystal interface. Amorphous material will have rubbery or glassy material behavior depending upon its glass transition temperature.

The glass transition temperature is an important characterizing parameter for amorphous polymers. The glass transition temperature is defined as the temperature at which a liquid or rubbery material becomes a glass, and hence, Tg can only correctly be measured on cooling [19–21].

2.1.5   Surface Quality

The critical factors that determine the surface quality are low gloss appearance along with good scratch and abrasion resistance [22].

2.1.6   Viscosity

Plastic material viscosity is influenced by four main factors: shear rate, temperature, molecular weight, and pressure [23–25].

2.2   Polyethylene

Polyethylene is used to produce products by injection molding, extrusion, and thermoforming operations. Polyethylene has toughness and stiffness and hence is used to produce hollow parts with rotational molding. It has many advantages such as chemical, electrical, and water resistance. However, it undergoes environmental stress cracking which gives brittle failure. Stress cracking resistance increases with decrease in density. Ductile failure occurs at very low temperature.

Although ductile failure occurs with PE, it is generally stable and undergoes photo oxidation. Polyethylene is the largest volume polymer used in the plastics industry due to the inexpensive nature of its ethylene monomers [26]. Polyethylene is a homopolymer of ethylene or copolymers of ethylene with other monomer contents. It is described by its density.

Normally polyethylene is insoluble in all solvents. It is also resistant to being soluble in polar solvents, vegetable oils, water, alkalis and most concentrated acid including hydrofluoric acid (HF) at room temperature.

Low density polyethylene (LDPE) is soft, flexible, and unbreakable [27–28]. LDPE is a most useful and widely used plastic especially in dispensing bottles or wash bottles. LDPE is ideally suited for a wide range of molded laboratory apparatus including wash bottles, pipette washing equipment, general purpose tubing, bags, and small tanks.

Low density polyethylene will dissolve slowly above 50°C in hydrocarbons, chlorinated hydrocarbons [29], higher aliphatic esters, and aliphatic ketones [30].

High density polyethylene (HDPE) has balanced mechanical properties with good chemical resistance. It has better barrier properties against gases and vapors. It is also harder and stiffer. The structure of HDPE is described as semi-crystalline [31]. LDPE is soft and flexible [32, 33]. High density polyethylene has been identified as a primary material for solid waste minimization and recycling in blow molding.

As the molecular weight, crystalline structure, and density change from the effect of multiple extrusion passes, environmental stress crack resistance (ESCR) also changes. This property is particularly important to HDPE blow molding resins, because changes in ESCR could have an effect on the shelf life of a bottler and on the type of fluids it can contain. The definition of ESCR is given as the promotion of cracking in the presence of an environment and a stress [34]. While the mechanism of environmental stress cracking (ESC) is not clearly understood, it is agreed that an aggressive medium tends to accelerate stress cracking. One possible explanation given by Williams [34] suggests that, under stress, a microvoided zone develops at the tip of a crack. This porous region tends to have a large surface area to material volume ratio which allows for rapid diffusion of the media into the thin ligaments. [35].

HDPE is an alternative for pipes to supply drinking water and gas and is produced by extrusion process. Polyethylene (PE) as tubes is also used in drainage schemes, either for acid waste, effluent, or laboratory discharge. Molded PE fittings are used as waste traps [36]. High density polyethylene lends itself particularly well to blow molding, e.g., for bottles, cutting boards, dipping baskets, dippers, trays, and containers.

Polyethylene accounts for more than 85% of the volume of rotomolding. Polyethylene has acceptable processing characteristics and good thermal properties during rotomolding and also good mechanical properties [37–55].

Ultra high molecular weight polyethylene is difficult to process by conventional processing due to higher melting point and viscosity. The extremely high molecular weight of ultra high molecular weight polyethylene (UHMWPE) products imparts properties to the plastic that limit the processing of the virgin material powder to ram extrusion and compression molding process.

Table 2.1 Properties of polyethylene (PE).

Properties	Value	Unit	Ref.
Specific gravity	0.92–0.94	–	[61, 62]
Melting temperature	110	°C	[61]
Specific volume	1.295	cm^3/g	[60]
Service temperature	55–70	°C	[61]
Processing temperature	205–260	°C	–
Transition temperature	105	°C	–

Polyethylene is used in rotomolding because of its good thermal stability, mechanical properties, and processing nature [41, 56 , 57].

HDPE combines low cost, excellent processability, outstanding resistance to many chemicals, and good toughness. HDPE also offers outstanding processing flexibility, affording, for example, the ability to mold-in features such as handles and threads [58].

Polyethylene can be differentiated by differential scanning calorimeter (DSC) [59].

2.3 Polypropylene (PP)

Polypropylene is stiffer and semi-crystalline in nature. It has high heat distortion temperature. It offers excellent chemical and environmental stress cracking resistance and surface hardness. Polypropylene is a low cost material for packaging or gasoline tanks that meets technical requirements such as barrier and mechanical properties [63].

PP products can be manufactured by any standard processing methods such as injection molding, extrusion, blow molding, thermoforming, etc. Physical properties of PP allow a great deal of flexibility in product. It can be easy to color and has living hinge capability. It has much better impact strength. PP has a versatile balance of stiffness, strength, and toughness which permits material substitution in many cases.

Polypropylene offers excellent chemical resistance, environmental stress cracking, and surface hardness. PP has high heat distortion temperature. It is autoclavable and requires higher melt flow index (MFI) to produce rotational molded products from grades suitable for injection molding polypropylene. Polypropylene has high heat distortion temperature. However, it cannot prevent the material from thermo-oxidative degradation during the rotational molding cycle. Polypropylene possesses other desirable properties to rotomold; because of brittleness, it has been excluded from the rotational molding process. There must be a balance with stiffness; polypropylene has been blended, copolymerized to produce the desirable properties by rotational molding.

PP has lower specific gravity than other plastic materials. Having broad resistance to organic chemical ingredients, they are used in consumer products. The total environmental impact of PP and other thermoplastic materials is less than traditional materials in life-cycle analysis. Commercial PP is a complex mixture of varying amounts of isotactic, syndiotactic, and atatic polymers with a given MWD.

The poor transparency and brittleness of PP restricts its application in the field of medical and personal care [64]. Blending isotactic PP with styrene/ethylene-butylene/styrene (SEBS) exhibits remarkable transparency with haze value as low as 6% along with a good percentage of elongation and processability.

A major advantage is polypropylene's higher temperature resistance, which makes PP particularly suitable for items such as trays, funnels, pails, bottles, carboys and instrument jars that have to be sterilized frequently for use in a clinical environment. Polypropylene is a translucent material with excellent mechanical properties. PP has a wide range of applications such as packaging, fibers, automobile industry, non-durable goods and in building construction.

In PP, efforts have been made to improve these two properties of strength and toughness [65–67].

Table 2.2 Properties of polypropylene (PP).

Properties	Value	Unit	Ref.
Melting point	439 (166)	K (°C)	[69]
Density	0.905	g/cm^3	[69]
Crystallization temperature	382 (109)	K (°C)	[69]
Specific volume	1.31–1.32	cm^3/g	[68]
Glass transition temperature	−17	°C	[70]
Processing temperature	200–230	°C	–
Degradation temperature (max)	479	°C	[70]

2.4   Polystyrene

Polystyrene has better processing ability with reduced softening temperature. Polystyrene responds less to shear history variations, in general. Commercial polystyrene is made from free radical polymerization. It has low density, transparency, high modulus, low cost, and easy processing. However, it is brittle in nature due to its structure. Heat of crystallization reduces the cycle time in injection molding. PS has low shrinkage values with high dimensional stability during molding and forming operations. Mineral oil reduces the viscosity of polystyrene. It is suitable for manufacturing inexpensive thin-walled parts, such as disposable dinnerware and toys. Using injection molding technique, it can be possible to produce medical ware, and using injection blow molding techniques it is possible to produce bottles, and extruded food packaging.

Table 2.3 Properties of polystyrene (PS).

Properties	Value	Unit	Ref.
Specific gravity	1.05	–	[74, 76]
Specific volume	1.02–1.04	cm^3/g	[71, 72]
Upper use temperature	105	°C	–
Processing temperature	200–230	°C	–
Glass transition temperature	98	°C	[76]
Transition temperature	27	°C	[75]
Degradation temperature max	398	°C	–
Melting temperature	210–220	°C	[73]
Vicat softening point	108	°C	–

2.5   Polyvinylchloride (PVC)

PVC in rigid form is used as plumbing and construction materials. With flexibility, PVC has transparency and impact

Table 2.4 Properties of poly(vinylchloride) (PVC).

Properties	Value	Unit	Ref.
Specific gravity	1.34–1.4	–	[61, 81, 87]
Service temperature	65	°C	[61]
Glass transition temperature Tg	81	°C	[80, 82, 83, 87]
Processing temperature	190–200	°C	–
Deflection temperature	75	°C	[84]
Upper use temperature	60–90	°C	[86]
Transition temperature	165	°C	[84]
Vicat softening temperature	70–80	°C	[85]

properties. These properties help its use in the field of medical and personal care [77]. Flexible PVC is also used in clothing and upholstery. Heat resistant PVC and ABS blends can improve the extensive resistance of plasticizer in isooctane and reduce weight loss in oleic acid, but reduces the flow characteristics [78].

Polyvinylchloride is generally transparent with a bluish tint. It is attacked by many organic solvents but it has a very good resistance to oils and it has a low permeability to gases.

PVC has outstanding resistance to both atmospheric and chemical resistance has been largely responsible for the growing importance in the chemical industry [79]. It has better resistance to inorganic chemicals, in addition to improved resistance to certain organic chemicals and somewhat lower permeability to gases [61]. It has excellent dielectric properties. PVC has proved an effective polymer for use in electrical wire and cable insulation core and sheathing. PVC has high chemical and abrasion resistance, and is widely used in durable applications, e.g., for pipes, window profiles, house siding, wire cable insulation, and flooring.

PVC usage in rigid packaging, including containers and bottles, has declined sharply, affecting the heat stabilizer industry to some extent, but it is still widely used in flexible packaging. However, PET has achieved market saturation in certain countries in the bottle and container market.

2.6   Engineering Plastics

Engineering plastics are thermoplastics with specific strength greater than usual metallic materials and relatively low cost. Hence, engineering plastics have recently replaced metallic materials in many different machine parts [88].

Engineering plastics are often used [89–91] and provide better properties at an economical cost. Engineering plastics are:

1. the family of nylon
2. polycarbonate
3. polyphenylene oxide
4. polyacetal
5. engineering grade of ABS
6. polysulphone and
7. polyphenylene sulphide. [91]

Engineering plastics have good mechanical properties and are frequently used material for various machine elements. During processing, hygroscopic materials such as nylon, ABS, etc., require pre-drying to avoid initial problems. Plastics are non-conductors of heat. Hence, they generate frictional heat that helps the material to melt faster during the processing operation. Despite their large diffusion, temperature range is controlled by the type of materal and their use.

Engineering plastics perform for prolonged use in structural applications over a wide range of temperature under mechanical stress and in difficult chemical and physical environments. These properties are either advantages or disadvantages, depending on use.

Engineering applications for plastics include mechanical units under stress, low friction components, heat and chemical resistant units, electrical parts, housings, high light transmission applications, building construction functions, and many miscellaneous uses.

Engineering plastics can be flexible and offer advantages such as transparency, self-lubrication, economy in fabricating, and decorating. Plastics are electrical non-conductors and thermo insulators. Plastics are considered to be competitive primarily with metals. Compared to metals, plastics are easier to fabricate and low cost. Plastics can be pigmented in a wide variety of colors.

Higher molecular weight usually gives stronger physical properties but can be harder to process. Numerous emerging technologies exploit the characteristics and/or properties of polymer [92]. Thermoplastics are the most widely used materials from non-critical packaging products to very demanding technical parts.

Resin is applied usually to the long chain of repeated units in the polymeric materials. This may be linear, branched, or cross-linked. Other constituents that contribute along with plastics as additives are plasticizers which improve the flexibility of the plastic, colored pigments, antioxidants, UV stabilizers, lubricants, mold-release agents, and sometimes biocides to inhibit microbiological deterioration. Additives influence the physical properties and resistance to deterioration in plastics.

Processes are influenced by the thermal characteristics of plastics. Melt temperature, glass transition temperature, and decomposition temperature are the main important thermal properties to be considered while processing.

2.6.1 Acrylonitrile Butadiene Styrene (ABS)

ABS is an engineering resin used extensively in industry owing to its good mechanical and processing properties.

To enhance its tensile strength, impact toughness, and stiffness further to reduce production cost, it is usually filled with rubber particles [93], or rigid inorganic particles, such as calcium carbonate [94], kaolin and glass bead [95], and talcum powder [96].

ABS possesses outstanding impact strength and high mechanical strength, which makes it so suitable for tough consumer products. Additionally, ABS has good dimensional stability and electrical insulating properties. ABS is a "terpolymer" with limited outdoor application. The durability of acrylonitrile–butadiene–styrene (ABS) polymers is important in many applications and depends on composition, processing, and operating conditions.

ABS is a well-known and widely used rigid engineering polymer. The mechanical properties of ABS are critical to its proper functioning in a given application, such as a medical device. It is, therefore, important to retain those properties during processing, fabrication, and use [97]. ABS is mainly used in appliance housings, canoes, computer keyboards, pipes and fittings, telephone and mobile housings, etc. It can be processed with normal processing methods of injection, blow molding, extrusion, and thermoforming. It exhibits viscoelastic behavior in both the melt and solid states. ABS has high impact strength over a wide temperature, with very good rigidity and toughness.

Acrylonitrile-butadiene-styrene terpolymer is one of the engineering plastics most frequently used as outer casings for computer equipment such as monitors, keyboards, and other similar components [98].

2.6.2 Polymethylmethacrylate (PMMA)

PMMA is superior from the optical point of view due to its high degree of transparency. It has good mechanical properties. However, it is slightly brittle in nature. PMMA is a rigid plastic and resistant to inorganic acids and alkalis, but is

Table 2.5 Properties of acrylonitrile-butadiene-styrene (ABS).

Properties	Value	Unit	Ref.
Specific gravity	1.04	–	[99]
Processing temperature	210–230	°C	–
Glass transition temperature Tg	95	°C	[99]
Deflection temperature	80–90	°C	–
Transition temperature	105	°C	–
Upper use temperature	60	°C	–

attacked by a wide range of organic solvents. In sheet form, it may be cemented to produce tanks, trays, racks, etc.

PMMA is light weight with low scratch resistance and variation of strength with temperature. It will not absorb the radiant heat of the sun. However, in opaque form, the surface temperature of PMMA rises. Increase in surface temperature lowers its physical properties and resistance to cold flow. It can be used to a limited degree in internal non-stressed parts, such as knobs, instrumental dial faces, and placards, where their high rate of change of strength with temperature is of little importance.

Poly(methyl methacrylate) is popular and well known, having been studied during several decades because of its broad practical applications. Owing to its excellent properties, PMMA is used in architecture, industry, motorization (as constructional materials and organic glasses in buildings, cars, ships, aircraft), in agriculture, medicine, pharmacy as well as in the textile, paper, and paint industries [30, 100].

PMMA can be used to manufacture the products either by injection molding of granulated material or extrusion of sheet from which it will be heated until soft and stretched over forms of the desired shape or pressed into the desired shape in heated dies. Both processes need cooling before the parts are removed from the die in order to prevent distortion.

Other uses for PMMA include in adhesives, automotive signal lights, lenses, light fittings, medallions, neon signs, and protective coatings because of the excellent, optical (clarity), physical and mechanical (dimensional stability with high modulus) properties.

Table 2.6 Properties of PMMA.

Properties	Value	Unit	Ref.
Specific gravity	1.18–1.246	–	[101–106], [110]
Melting temperature	160	°C	[107]
Service temperature	70	°C	–
Glass transition temperature Tg	85–105	°C	[101, 106, 108, 109]
Processing temperature	220–250	°C	–
Deflection temperature	100	°C	–
Vicat softening point	82–110	°C	[108]

2.6.3 Nylon

Nylon is among the best known of all engineering plastics. It has high strength and good resistance to wear, fatigue, heat, and chemicals. It can be processed by injection molding, extrusion, blow molding, rotational molding, and thermoforming. Nylon reinforced with glass fiber is used to manufacture precision parts with high molecular performance by injection molding. Even with high fiber content, nylon is easy to inject and has high productivity.

Major uses include in electrical connectors, gears, bearings, cable ties, fishing line, automotive valve covers, oil pans, sports and exercise equipment. Commercially, nylon is commonly used in the production of tire cords, rope, belts, filter cloths, sports equipment, and bristles. It is particularly useful when machined into bearings, gears, rollers and thread guides.

Nylon is a crystalline polymer with good mechanical and thermal properties and is an important material used in high-performance molded applications such as radiator end tank, intake manifold, engine cover, timing belt cover, and various types of automobile parts like automatic transmission thrust washer, wiring harness clip, and pump rotor guide [111, 112].

Nylons are an attractive class of engineering polymers due to their excellent strength and stiffness, low friction, and chemical and wear resistance [111]. However, they are highly notch sensitive, that is, they are often ductile in the unnotched state, but fail in a brittle manner when notched. In addition, nylon tends to be brittle at low temperatures and under severe loading conditions. Fortunately, the inherent chemical functionality of nylons makes them an attractive candidate for modification. [113–124].

Table 2.7 Physical properties of nylon.

Properties	Value	Unit	Ref.
Specific gravity	1.1	–	[61]
Melt density (r)	1000	Kg/m^3	[125]
Glass transition temperature $Tg = Tm$ (K)	318 (45)	K(°C)	[125]
Service temperature (max.)	120	°C	[61]
Melt temperature	270–320	°C	–
Processing temperature	270–290	°C	–

2.6.4 Polyethyleneterephthalate (PET)

Poly(ethylene terephthalate) (PET) has taken a central position under engineering plastics and is commonly known as polyester. As fiber, bottle, and film material or as matrix for glass-reinforced plastics, PET has found a wide field of application. It is distinguished by very good processability, low shrinkage, low water content, and barrier properties. Limitations arise from relatively low glass transition temperatures resulting in reduced thermal stability. To overcome this problem polyesters are often reinforced by glass fibers and/or cross-linked by various methods. Such materials are being used in electrical applications [126].

Polyesters are one of the most versatile classes of polymers ever produced and nowadays, aliphatic polyesters are one of the most promising alternatives to commodity plastics. They can be obtained by a wide range of reactions, the most important being the polyesterifications between diacids and diols or their derivatives. Depending on the nature of alkyl groups in both diacid and diol, an enormous wide variety of structures, architectures, properties and applications are available [127]. The low melting points of aliphatic polyesters have prevented their wide usage as polymeric materials for a long time. However, because of their characteristic biodegradability and the ongoing environmental concerns, aliphatic polyesters are now in the middle of the spotlight [128]. Polyethylene is the dominant packaging polymer, used for high-volume supermarket bags, food packaging, and rubbish sacks. HDPE is the most important resin for rigid packaging in Europe, but polystyrene is widely used in packaging sheet, as is PVC to a lesser extent. Polypropylene is preferred for the more specialized role of packaging industrial goods.

Table 2.8 Physical properties of poly(ethylene terephthalate) (PET).

Properties	Value	Unit	Ref.
Specific gravity	1.35	–	–
Melting temperature Tm	256	°C	[129]
Glass transition temperature Tg	76	°C	[129]
Service temperature (max.)	100	°C	–
Processing temperature	280–300	°C	–
Transition temperature	220	°C	–

2.6.5   Polycarbonate (PC)

Polycarbonate is attractive, particularly due to its impact resistance and good strength at elevated temperatures. It also has exceptional toughness and is formable to deep compound contour. However, it has disadvantages of poor optical quality due to inherent limitation of the extrusion process, poor abrasion and chemical resistance. It may bubble at elevated temperature and the surface is degraded by outdoor weathering.

Polycarbonate is a true thermoplastic and, unlike the stretch altered PMMA, does not undergo molecular orientation when heated to its softening point.

PC is an important engineering thermoplastic having unique properties like transparency, toughness, thermal stability, and dimensional stability. These properties allow uses in many applications like compact discs, riot shields, vandal-proof glazing, baby feeding bottles, electrical components, safety helmets, and headlamp lenses.

Polycarbonate possesses extraordinarily good dimensional stability with a high impact strength which is maintained over a wide range of temperature. This property makes PC an ideal material for the manufacture of safety shields, vacuum desiccators, and centrifuge tubes.

Polycarbonate's continuous maximum working temperature 130°C makes PC to be autoclaved. Repeated autoclaving makes PC lose some mechanical strength and should not be then used for vacuum applications.

2.6.6   Polyether Ether Ketone (PEEK)

PEEK is a semi-crystalline thermoplastic polymer (typically 35%) with outstanding mechanical properties, high melting point, and good resistance to strong acids. The high performance PEEK was first prepared by Bonner in 1962 [131].

PEEK can be readily injection molded with most standard reciprocating screw injection molding machines. Complex

Table 2.9 Physical properties of polycarbonate (PC).

Properties	Value	Unit	Ref.
Specific gravity	1.19	–	[62]
Melting temperature Tm	270	°C	–
Glass transition temperature Tg	150	°C	[130]
Vicat softening point	430 (157)	K(°C)	[130]
Processing temperature	280–305	°C	–
Transition temperature	144	°C	–

high performance components can be readily mass produced without need for annealing or conventional machining. However, due to the higher melting point of PEEK, certain process variables and design need to be considered.

PEEK can be readily molded by using standard reciprocating screw injection molding machines. Complex high performance components can be readily produced without any annealing or conventional machining. However, due to high melting temperature, design and process variables need to be considered. Optimal processing parameters could help to solve most quality control problems.

Most standard reciprocating screw injection molding machines are capable of molding PEEK polymer and compounds. Complex high performance components can be readily mass produced without need for annealing or conventional machining. PEEK polymer compounds based on PEEK polymer can be readily injection molded. However, due to the high melting temperature, certain design and process variables need to be considered. The configuration of the processing parameters for the injection molding process is the focus since an optimal processing parameter design could help solve most quality control problems.

The adjustment of the processing parameters is often done considering the mold cavity design and its size, the properties of plastic materials, and the defects of the molding product, etc. Such tasks require accumulated data and experience from a large number of tests and experiments to clarify the causes of product defects. They also entail a time and effort consuming process.

Table 2.10 Physical properties of polyether ether ketone (PEEK).

Properties	Value	Unit	Ref.
Specific gravity	1.3	–	[133]
Melting temperature Tm	335	°C	[132]
Glass transition temperature Tg	145	°C	[132]
Degree of crystallinity	30–35	%	[133]

2.6.7   Polytetrafluoroethylene (PTFE)

Polytetrafluoroethylene (PTFE) shows a remarkable chemical resistance and it is insoluble in all known solvents. It is attacked only by molten alkali metals and by fluorine at high temperatures. PTFE is incombustible and may be used up to 260°C (300°C for short periods). PTFE's coefficient of friction is extremely low and it shows the effect of self-lubrication with constant mechanical properties which makes it particularly suitable for bearings, joints, O-rings, stirring bars, hazardous materials bottle pourers, syringes, crucibles, evaporating dishes, etc. PTFE may be shaped by compression and sintering into bottles and beakers where its chemical stability and non-wettability make it suitable for use in extreme circumstances.

PTFE, commonly known as Teflon, is resistant to all chemicals except molten alkali metals. Even though very expensive, it is an attractive material for critical corrosion resistant application. It can be used at temperature up to 287°C. It is difficult to fabricate and cannot be molded by the usually accepted methods. PTFE tubing is made by making a pre-form under pressure at ordinary temperatures and then sintering this at about 354°C.

Table 2.11 Properties of polytetrafluoroethylene (PTFE).

Properties	Value	Unit	Ref.
Specific gravity	2.18	–	[134]
Melting temperature Tm	327	°C	[135]
Glass transition temperature Tg	127	°C	[135]
Upper use temperature (max.)	230–260	°C	[135]
Deflection temperature	60	°C	[136]
Transition temperature	80	°C	–

2.6.8   Polyacetal (POM)

Acetal plastics became commercially available in 1960 and are commonly known as polyacetal (POM). Acetal is a by-product of a two-step reaction between an alcohol and an aldehyde, formed by polymerizing anhydrous formaldehyde to form an oxymethlylene chain. Polyacetal is a popular and versatile crystalline engineering polymer. It is very strong and exhibits good chemical resistance. Also it is a tough material with good dimensional stability and a low coefficient of friction. Polyacetal machines well and is relatively easy to process. Typical applications include gears, springs, plates, bushings, and housings. It is formaldehyde based polymer and care should be taken when processing; formaldehyde released is a skin and eye irritant.

Table 2.12 Physical properties of polyacetal (POM).

Properties	Value	Unit	Ref.
Specific gravity	1.42	–	[137]
Melting temperature Tm	448(175)	K(°C)	[137]
Deflection temperature	409(136)	K(°C)	[137]
Glass transition temperature (Tg)	198(−75)	K(°C)	[138]

2.6.9 Polyvinylidene Fluoride (PVDF)

Polyvinylidene fluoride (PVDF) is a valuable thermoplastic resin due to its good mechanical strength, rigidity and toughness as well as its wide range of service temperatures and high chemical and temperature resistance [139–140]. This property profile opens applications for PVDF for example in pipes, fittings, pumps, and plenum wires [139, 141].

PVDF offers remarkable chemical inertness and excellent thermo-mechanical properties with very high resistance to ageing. It is used in aggressive environments where temperatures can vary between −40°C and +150°C in a variety of applications such as pipes, valves, pumps, measuring devices, internal conduits of chimneys where it affords better environmental protection, reactors, tanks, heat exchangers, etc. It is very easy to process with conventional processing techniques such as extrusion, injection molding, rotational molding, etc.

The high fluidity of compounds allows the injection molding cycle to be shortened substantially. With reinforcing mechanical properties, the appearance of the finished parts dramatically improves during processing.

PVDF offers very good impermeability to water and water vapor, gases, and odors, in both dry and wet environments. It will be used for extruding and co-extruding films and sheets.

Table 2.13 Physical properties of polyvinylidene fluoride (PVDF).

Properties	Value	Unit	Ref.
Specific gravity	1.75–1.78	–	[143]
Melting temperature Tm	451(178)	K(°C)	[142]
Glass transition temperature Tg	238(−35)	K(°C)	[142]
Melting point	173.0	°C	[144]

2.6.10 Polyphenylene Sulfide (PPS)

PPS is among the engineering polymers reinforced with glass fibers and/or mineral filers. It is primarily intended for injection molded parts requiring high resistance to temperature, fire, and chemicals.

Table 2.14 Properties of polyphenylene sulfide (PPS).

Properties	Value	Unit	Ref.
Specific gravity	1.32–1.43	–	[146]
Melting temperature Tm	280–320	°C	[145, 146, 148]
Glass transition temperature Tg	83	°C	[149]
Upper use temperature (max.)	200	°C	–
Deflection temperature	136.7	°C	[147, 150]

2.7 Advantages

The advantages and growth of plastics may be attributed to the properties which include:

1. Lower density than metals and ceramics
2. Tailoring and specific needs with high versatility
3. Finishing, painting or polishing not required
4. Excellent thermal and electrical properties
5. Reduction in fuel consumption during transportation
6. Provide safety and hygiene properties for food packaging

2.8 Fundamentals

- Plastics are man made synthetic materials
- Plastics can be formulated or designed with technical innovation to offer properties to suit particular applications

- Plastics materials in commercial use available with more than 50 different families with subtype and variations
- Plastics have a very low thermal conductivity and melt slowly
- Melt behavior of a plastic depends on the melt rheological properties of the polymer
- The characteristics of the material will be derived from the material's density, melt index, molecular weight, viscosity number, and the crystalline melting point
- Molecular weight distribution is the most important property of plastics

References

1. Siangchaew, K., and Libera, M., Energy-Loss Measurements of Polymer Microstructure and Polymer Interfaces: Issues and Opportunities, *Microsc. Microanal.* 3, 530–539, 1997.
2. Alauddin, M., Choudhury, I.A., El Baradie, M.A., Hashmi, M.S.J., *Journal of Materials Processing Technology* 54 (1995) 40–46.
3. Jacquet, J.P., Plastic optical fibre applications for lightening of airports and building, *Proc. of SPIE*, The Intl. Society for Optical Eng., 1592 (1991) 165–172.
4. Weber, A., Plastics in automotive engineering use and reuse, *J. Materials Design*, 12 (1991) 199–208.
5. Leaversuch, D.R., Room air conditioner with no metal bending, *J. Modern Plastics*, 68 (1991) 48–50.
6. Allbee, N., Plastics in the medical world, *J. Plastics Compounding*, 12 (1989).
7. Anzano, J., Casanova, M.E., Bermüdez, M.S., Lasheras, R.J., Rapid characterization of plastics using laser-induced plasma spectroscopy (LIPS), *Polymer Testing* 25 (2006) 623–627.
8. Chum, P.S., Swogger, K.W., Olefin polymer technologies – History and recent progress at The Dow Chemical Company, *Progress in Polymer Science* 33 (2008) 797–819.
9. Dutor, J.Y., *J. Materials Design*, 7 (1) (1986) 14.
10. Migler, K.B., Son, Y., Flynn, K., Extensional deformation, cohesive failure, and boundary conditions during sharkskin melt fracture, *J. Rheol.* 46 (2002) 383–400.
11. DeGarmo, E.P., Black, J.T. and Kohser, R.A., (1988), *Materials and Processes in Manufacturing*, 7th ed., Macmillan, London.

12. Geil, P.H., *J. Chem. Educ.* 1981, 58, 879.
13. Schepper, B., Ewering, J., *Plastverarbeiter* (2003), 54(12), 40–41.
14. Hempenius, M.A., Zoetelief, W.F., Gauthier, M., Möller, M., Melt Rheology of Arborescent Graft Polystyrenes, *Macromolecules* 1998, 31, 2299–2304.
15. Shroff, R., and Shida, M., Effect of molecular weight and molecular weight distribution on elasticity of polymer melts, *ANTEC* (1977).
16. Christensen, R.E., and Cheng, C.Y., *Plast. Eng.* June 1991.
17. Schott, H., and Kaghan, W.S., *J. Appl. Polym. Sci.* 14, 175, 1961.
18. Zahavich, A.T.P., Latto, B., Takacs, E., Vlachopolulos, J., The effect of multiple extrusion passes during recycling of high density polyethylene, *Polym Techn* 16:11–24, 1997.
19. McKenna, G.B., and Simon, S.L., *Handbook of Thermal Analysis and Calorimetry, Applications to Polymers and Plastics*, Vol. 3, S.Z.D. Cheng (Ed.), Elsevier Science, pp. 49–109 (2002).
20. Plazek, D.J., and Ngai, K.L., *Physical Properties of Polymers Handbook*, Mark, J.E. (Ed.), AIP, Woodbury, NY, pp. 139–159 (1996).
21. McKenna, G.B., *Comprehensive Polymer Science, Polymer Properties*, Vol. 2, Booth, C., and Price, C., (Eds.), Pergamon, Oxford, Ch. 2, (1989).
22. Rogers, S., Walia, P.,Van Riel, N., Van Dun, J., Traugott, T., *Annual Technical Conference – Society of Plastics Engineers* (2009), 67th 1037–1041.
23. Barnes, H.A., Hutton, J.F., and Walters, K., *An introduction to rheology*, Elsevier, Oxford, U.K. 1989.
24. Duncan, R.E., and Zimmerman, A.B. *Rotational molding of high density polyethylene powders, Society of plastics Engineers*, Regional Technical conference, p. 50, (March 1969).
25. Cogswell, F.N., *Plastics and polymers*, February 1973, p39.
26. Vasile, C., *Handbook of Polyolefins*, 2nd ed.; Marcel Dekker: New York, 2000.
27. Alger, M., *Polymer Science Dictionary*, 2nd ed., Chapman and Hall, 2nd edn., Chapman and Hall, 1997.
28. Tasdemir, Munir; Biltekin, Hasan; Caneba, Gerald T., Preparation and characterization of LDPE and PP-wood fiber composites, *Journal of Applied Polymer Science* (2009), 112(5), 3095–3102.
29. Van Krevelen, D.W., and Hoftzer, P.J., (1976) *Properties of Polymers*, 2nd ed., Elsevier, Amsterdam, 5.
30. Bloch, D.R., *Polymer Handbook.*, 4th ed., (eds. Brandrup, N.J., Immergut, E.H., and Grulke, E.A.), John Wiley & Sons, 1999, vol VII, 500.
31. Allen, N.S., *Degradation and Stabilization of Polyolefins*, Applied Science, New York, 1983.
32. Alger, M., *Polymer Science Dictionary*, 2nd ed., Chapman and Hall, 1997.
33. Myers, J.H., *J. Plas. Film and Shtg.*, 1:250–257 (1985).
34. Williams, J.G., *Fracture mechanics of polymers*, John Wiley & Sons, Toronto, 1984.

35. Wang, H., Cao, B., Jen, C.K., Nguyen, K.T., and Viens, M., *Polym. Eng. Sci.*, 1997, vol 37, 363.
36. Evans, V., *Corrosion Technology* April 1956, p112.
37. Wigotsky, V., *Plastics Engineering*. February 1998, pp. 18–23.
38. Bellehumeur, C.T., Bisaria, M., and Vlachopoulos, J., "Study of Sintering in Rotational Molding," SPE *ANTEC Tech Papers* II, 41, 1973 (1995).
39. Crawford RJ, editor. Rotational moulding of plastics. *Polymer engineering series*. 2nd ed. New York: Research Studies Press; 1996.
40. Throne, J.L., *Plastics Process Engineering*, pp. 579–614, Marcel Dekker, New York (1979).
41. Coughlan, J.J. and Hug, D.P., *Encycl. Polym. Sci. Eng.*, 2nd edn. (ed. Kroschwitz J.L.) John Wiley & Sons, 1986, vol 6, 490.
42. Bisaria, M.K., Takacs, E., Bellehumeur, C.T., and Vlachopoulos, J., *Rotation*, 3 (4), 12 (1994).
43. Lontz, J.F., *Fundamental Phenomena in Material Sciences*, Bonis, L.J., and Hanser, H.H., Editors, Plenum Press, New York (1964).
44. Rao, M.A., and Throne, J.L., *Polym Eng. Sci*, 12, 237 (1972).
45. Throne, J.L., *Polym Eng. Sci*, 18, 257 (1976).
46. Throne, J.L., and Sohn, M.S., *Adv. Polym Technol.*, 9 (3), p181 (1989).
47. Oliveira, M.J., Cramez, M.C., Crawford, R.J., Structure – properties relationship in rotationally moulded polyethylene. *J Mater Sci* 1996; 31(9): 2227–40.
48. Crawford, R.J., and Nugent, P.J., *Plast Rubber Process Appl.*, 11, 107 (1989).
49. Iwakura, K, Ohta Y., Chen C.H., and White J.L., *Int. Polym Process, IV*, 163 (1989).
50. Chen C.H., White J.L., and Ohta Y., *Polym Eng. Sci*, 30, 1523 (1990).
51. Xu, L., and Crawford. R.J., *Plast. Rubber Compos. Process Appl.*, 21, 257 (1994).
52. Bawiskar, S., and White, J.L., *Polym Eng. Sci*, 34, 815 (1994).
53. Bawiskar, S., and White, J.L., *Int. Polym Process, X*, 62 (1995).
54. Takacs, E., Bellehumeur, C.T., and Vlachopoulos, J., *Rotation*, 5 (3), 17 (1996).
55. Bellehumeur, C.T., Kontopoulou, M., and Vlachopoulos, J., *RheoL Acta*, 37 (3),2 70 (1998).
56. Pennings, A.J., Vanderhooft, R.J., Postema, A.R., Hoogsteen W., Tenbrinke G., High-speed gelspinning of ultra-high molecular weight polyethylene. *Polym Bull* 1986;16:167–174.
57. Kontopoulou, M, Takács, E, Bellehumeur, C.T., and Vlachopoulos. J, Plastics engineering, Feb 1998, pp. 29–31, 1998.
58. Luckey, Jr., S.G., Henshaw, J.M., Dewan, C., Eltanany, G.M., Teeters, D., Analysis of a blow molded HDPE bottle that failed by brittle fracture, *Engineering Failure Analysis*, 8, (2001), 361–370.

59. Wolf, B., Kenig, S., Klopstock, J., and Miltz, J., *J. Appl. Polym. Sci.*, 1996, vol 62, 1339.
60. Cudin, A., Chee, K.K., and Shaw, J.H., *J. Polym. Sci. C*, 415, 1970.
61. *Corrosion Technology*, April 1956, p. 106–108.
62. Grewell, David A.; Benatar, Avraham., Comparison of orbital and linear vibration welding of thermoplastics, *Polymer Engineering & Science* (2009), 49(7), 1410–1420.
63. McCormack., T., *Modern Plastics*, 1993, April, 77.
64. Ahmad, Zubair; Kumar, K. Dinesh; Saroop, Madhumita; Preschilla, Nisha; Biswas, Amit; Bellare, Jayesh R.; Bhowmick, Anil K., *Polymer Engineering & Science* (2010), 50(2), 331–341.
65. Zhang, Z., Yang, J.L., Friedrich K., *Polymer* 2004;45:3481–5.
66. Winter, D., Eisenbach, C.D., *Polymer* 2004;45:2507–15.
67. Chapleau, N., Mohanraj, J., Ajji, A., Ward, I.M., *Polymer*, 2005;46: 1956–66.
68. Lazár, M., Rado, R., and Rychlý, J. , *Adv. Polymer Sci.* 95, 149 (1990).
69. Joonsung, Yoon, Thomas, J., McCarthy., Lesser, Alan, J., Self-reinforcing thermoplastic composites, *ANTEC 2008*, 9.
70. Khlok, D., Deslanders,V., Prudhomme, J., *Macromolecules* 1976;9:809.
71. Chee, K.K., and Rudin, A., *I.E.C. Fundamentals*, 9, 177, 1970.
72. Covas, J.A., Costa, P., Polymer Testing 23 (2004) 763–773.
73. Maruselli, E., Demma, G., Driol, E., Nicolais, L., *Polymer*, 1979, vol 20, 571.
74. Schrader, D., *Polym. Handb.*, (eds. Brandrup. J., Immergut. E.H., and Grulke, E.A.), John Wiley and Sons, 4th ed., 1999, vol V, 91.
75. Boundy, R.H., and Boyer, R.F., (eds.), *Styrene, Its Polymers, Copolymers and Derivatives*, Reinhold Corp. (New York), 1952.
76. Fekete, Erika, Földes, Enikö, Béla Pukánszky, Béla, Effect of molecular interactions on the miscibility and structure of polymer blends, *European Polymer Journal*, 41 (2005) 727–736.
77. Ahmad, Zubair; Kumar, K. Dinesh; Saroop, Madhumita; Preschilla, Nisha; Biswas, Amit; Bellare, Jayesh R.; Bhowmick, Anil K., *Polymer Engineering & Science* (2010), 50(2), 331–341.
78. Jia, Yongpeng; Zhang, Suifa; Xu, Xiaofeng; Tao, Guoliang; Wu, Tieheng., *Xiandai Suliao Jiagong Yingyong* (2009), 21(3), 29–31.
79. *Corrosion Technlogy*, April 1956 p.115.
80. Dall'Asta, G., Meneghini, P., Gennaro, U., *Makromol Chem.*, 1972; 154:291.
81. Roff, W.J., and Scott, J.R., *Fibres, Films, Plastics and Rubbers*, Butterworths, 1971, 108.
82. Reding, F.P., Walter, E.R. and Welch, F.J., *J. Polym. Sci.*, 1962, vol 56, 225.
83. Keavney, J.J. and Eberlin, E.C., *J. Appl. Polym. Sci.*, 1960, vol 3 , 47.
84. Bockman, O.C., *Br. Plast.*, 1965, vol 38, 364.

85. Sweeting, O.J., Ed., *Sci. Technol. Polym. Films*, John Wiley & Sons, 1971, 307.
86. Whelan, A., *Polymer Technology Dictionary*, Chapman and Hall, 1994, 207.
87. Satov, D.V., Additives for wood-polymer composites, Editor(s): Niska, Kristiina Oksman, Sain, Mohini., *Wood-Polymer Composites conference* (2008) 23–40.
88. Hiroki, Endo, Etsuo, Marui, Fundamental studies on friction and wear of engineering plastics, *Industrial Lubrication and Tribology*, Volume 56, Number 5, 2004, 283–287.
89. Kommayer, Storage of plastics in outdoor silos, *J. German Plastics (Kunststoffe)*, 81 (1991) 12–14.
90. Suenaga, J., Fujita, E., and Marutani, T., Polymer blends of semi aromatic liquid crystal polymers with engineering plastics, *Japanese J. Polymer Sci. Tech.*, 48 (1991) 573–579.
91. McQuiston, H., Designing with engineering plastics, *J. Plastic Eng.*, (1980) 18–25.
92. Schwartz, S.S., Goodman, S.H., *Plastic Materials and Processes*, Van Nostrand Reinhold, New York, 1982.
93. Grancio, M.R., *Polym. Eng. Sci.*, (1972) 12: 213–223.
94. Tang, C.Y. and Liang, J.Z., *Proceedings of IMCC'2000*, pp. 45–46, Hong Kong, August 16–17, (2000).
95. Lavengood, Nicolais, R.E.L., and Narkis, M. (1973). *J. Appl. Polym. Sci.*, 17: 1173–1185.
96. Bigg, D.M., (1987). *Polym. Compos.*, 8: 115–122.
97. Blom, H., Yeh, R., Wojnarowski, R, and Ling, M, *Thermochimica Acta* 442 (2006) 64–66.
98. Anandhan, S., De, P.P.; De, S.K., Bhowmick, Anil, K., Bandyopadhyay, S., *Rubber Chemistry and Technology* (2003), 76(5), 1145–1163.
99. Boronat, T., Segui, V.J., Peydro, M.A., Reig, M.J., *Journal of materials processing technology* 209 (2009) 2735–2745.
100. Mark, H.F., In: Bikales NM, Overberger Ch G, Menges G, Kroschwitz JI, editors. *Encyclopedia of polymer science and engineering*, vol. 1. New York: John Wiley & Sons, 1985. p. 234.
101. Quach, L., Otsu, T.J., *Polym Sci.*, 1982;20:2501.
102. Pulos, Guillermo C., Knauss, Wolfgang, G., Constant stress intensity factors through closed-loop control, *International Journal of Fracture* (1993), 63(2), 101–12.
103. Boscher, F., Ten Brinke, G., Eshuis, A. and Challa, G., *Macromolecules*, 1982, vol 15, 1364.
104. Wiebking, H.E., *J. Vinyl Additive Technol.*, 12, 1, 37–40, 2006.
105. Hoff, E.A.W., Robinson, D.W., and Willbourn, A.H., *Polym. Sci.*, 1955, vol 18, 161.

106. Carrizales, C., Pelfrey, S., Rincon, R., Eubanks, T.M., Kuang, A., McClure, M., Bowlin, G.L., Macossay, J., *Polymers for Advanced Technologies*, 19, 2, 2008, 124–130.
107. Shetter, J.A., J. *Polym. Sci., Part B: Polym. Lett.*, 1962, vol 1, 209.
108. Harrington, M., *Handbook of Plastic Materials and Technology*, (ed. II. Rubin), Wiley Interscience, 1990, 355.
109. Halina Kaczmarek, Alina Kamińska, Alex van Herk, Photooxidative degradation of poly(alkyl methacrylate)s, *European Polymer Journal* 36 (2000) 767–777.
110. Kurtz, S.M., Devine., *Biomaterials*, 28 (2007) 4845–4869.
111. Kohan, M.I., *Nylon plastics handbook*, New York: Hanser/Gardner; 1995.
112. Lann, P.L., Derevjanik, T.S, Snyder, J.W.,Ward Jr, W.C., Nylon degradation with automatic transmission fluid, *Thermochim Acta* 2000;357–358:225–30.
113. Wu. S., *Polymer* 1985;26:1855.
114. Majumdar, B., Keskkula, H, Paul, D.R., *Polymer* 1994;35:1386.
115. Modic, M.J., Pottick, L.A., *Polym Eng Sci.*, 1993;33:819.
116. Takeda, Y., Paul, D.R., *J Polym Sci, Part B: Polym Phys* 1992;30:1273.
117. Oshinski, A.J., Keskkula, H., Paul, D.R., *Polymer* 1992;33:268.
118. Yamazaki, N., Higashi, F., and Kawabata, J., *J. Polym. Sci., Polym. Chem. Ed.*, 1974, vol 12, 2149.
119. Dijkstra, K., Gaymans, R.J., *Polym Commun.*, 1993;34:3313.
120. Borggreve, R.J.M., Gaymans, R.J., Schuijer, J., Ingen Housz, J.F., *Polymer* 1987;28:1489.
121. Borggreve, R.J.M., Gaymans, R.J., *Polymer* 1989;30:63.
122. Mancarella, C., Martuscelli, E., Palumbo, R., Ragosta, G., *Polym Eng Sci.*, 1984;24:48.
123. Lu, M., Keskkula, H., Paul, D.R., *J Appl Polym Sci.*, 1996;59:1467.
124. Misra, A., Sawhney, G., Kumar, R.A., *J Appl Polym Sci.*, 1993;50:1179.
125. Joo, Y.L., Sun, J., Smith, M.D., Armstrong, R.C., Brown, R.A., Ross, R.A., 2-D numerical analysis of non-isothermal melt spinning with and without phase transition. *J Non-Newt Fluid Mech.*, 2002; 102:37–70.
126. Böhme, F, Komber, K, Jafari, S.H., Synthesis and characterization of a novel unsaturated polyester based on poly(trimethylene terephthalate), *Polymer* 47 (2006) 1892–1898.
127. Rogers, M.E, Long, T.E., *Synthetic methods in step-growth polymers*, New York: John Wiley and Sons; 2003.
128. Nair, L.S., Laurencin, C.T., *Prog Polym Sci.* 2007;32(8–9):762–98.
129. Pingping, Z., Dezhu, M., Study on the double cold crystallization peaks of poly(ethyleneterephthalate) 3. The influence of the addition of calcium carbonate ($CaCO_3$), *European Polymer Journal*, 36 (2000) 2471–2475.

130. Mark, J.E., ed. *Physical Properties of Polymers Handbook,* AIP Press, Woodbury, N.Y., 1996.
131. Bonner, W.H., *U.S. Patent*, 3065205 (1962).
132. D'Aniello, C., Romano, G., Russo, R,.and Vittoria, V., *European Polymer Journal* 36 (2000) 1571–1577.
133. Blundell, D.J., and Newton, A.B., *Polymer*, 1991, vol 32, 309.
134. Kroschwitz. J.I., Editor, *Encycl. Polym. Sci. Eng.,* Wiley Interscience, 2nd edn., 1989, vol 3.
135. Waterman, N.A., and Ashby M.F., Editors, *The Materials Selector*, 2nd edn., Chapman and Hall, 2nd edn., 1997 vol 3.
136. Sperati, C.A., and Starkweather, H.W., *Adv. Polym. Sci.*, 1961, vol 2, 465.
137. Serle., A.G., *In Engineering Thermoplastics: Properties and Application*, edited by J.M., Marcel Dekker, New York, 1985, p. 151.
138. Aoki, Y., Nobuta, A., Chiba, A., and Kaneko, M., *Polymer J.* 2 (1971): 502.
139. Dominghaus, H., Eyerer, P., Elsner, P., Hirth, T., Editors, *Plastics and their characteristics*, 6th ed.; 2005.
140. Lovinger, A.J., *Developments in Crystalline Polymers*, 1982;1:195.
141. Salsmone, J.C., editor, *Polymeric Materials Encyclopedia*, Twelve Volume Set; 1996.
142. Kobayashi, M., Tashiro, K., and Tadokoro, H., *Macromolecules* 8(2) (1975):158.
143. Enns, J.B., and Simha, R., *J. Macromol. Sci., Phys.* B13(1) (1977):11–24.
144. Lovinger, A., *Science* 220 (1983): 1,115.
145. Cheng, S.Z.D., Pan, R., Bu, H.S., Cao, M.Y., and Wunderlich, B., *Makromol. Chem.*, 1988, vol 189, 1579.
146. Tabor, B.J., Magrè, E.P., and Boon, J., *Eur. Polym. J.*, 1971, vol 7, 1127.
147. Rigby, S.J., and Dew-Hughes, D., *Polymer*, 1974, vol 15, 639.
148. Könnecke, K., *J. Macromol. Sci., Phys.*, 1994, vol 33, 37.
149. Seo, K.H., Park., L.S., Baek., J.B. and Brostow, W., *Polymer*, 1993, vol 34, 2524.
150. Kourtides, D.A., and Parker, J.A., *Polym. Eng. Sci.*, 1978, vol 18, 855.

3

Plastics Additives

The plastics industry is a major consumer of additives. Additives are used to tailor the polymer properties for specific applications. Proper use of additives with plastics produces compounds with balanced and required properties. Selection of additive for a specific function should ensure no deterioration effect on other additives or the final product. Addition of different additives requires significant work to achieve the balanced properties in the final product.

An additive can be used in plastics for mold enhancement as well as antistatic property. It can also be used as a mold release agent and internal lubricant. Sometimes the same additive may be useful as a distribution agent for colorant for various plastics and to certain polymers as melt flow enhancer. One such example is glycerol mono stearate commonly known as GMS. Additives enhance surface, mechanical, chemical, and aesthetic properties.

Chemical interactions between the additives and plastics have been recognized and documented when the interaction was readily apparent [1–3]. Today, environmental and regulation issues change the plastics additive market. Demands are for less toxic, more efficient, and low cost additives.

Antioxidants, heat and light stabilizers, flame retardants, antistatic agents, and antimicrobials are property extenders and used in plastics to create more demanding applications. Improved processing and cost effectiveness is the driving force for innovative developments in additives. Additives are in demand particularly in packaging and consumer products.

The polymer and additives combination is called plastics. The properties of plastics can be modified with additives. The plastics are shaped to convert into useful products using processing techniques such as injection molding, extrusion, blow molding, etc. To have better molding, extruding, and other processes, an array of additives and fillers is included in the material before processing.

Additives are foreign substances which added to plastics melt strongly influence the crystal nature during the process. Many substances promote nucleation [4]; as a result significant changes in the morphology and the molecular structure occur. In many cases, additives result in different polymorphic forms being formed [5]. The influence of additives on the crystallization process which observed in the sluggish melt process, does not relate to the crystallization process that occurs during processing. In plastics extrusion processes, polymer resin is mixed with several additives at specific ratios to obtain the desired physical and mechanical properties during and after processing.

Plastics processing is made easy with the use of additives. Additives help to solve the processing problems. The problems occur from the inherent visco-elastic behavior of individual plastic types.

3.1   Antioxidants

Antioxidants prevent degradation of material during high temperature processing such as blow molding of bottles, and also increase its shelf life. Introducing antioxidants in plastics can exhibit the cross-linking ability and prevents the degradation of PE during rotational molding.

Efforts to modify polyethylene with antioxidants have been in progress since the 1960s [6–16].

Processing stabilization of polyethylene is usually done by a combination of phenolic and phosphorous antioxidants. A phosphate stabilizer used in the absence of a phenolic antioxidant imparts very low oxidative stability to polyethylene. When hindered phenols are used in combination with phosphites or phosphonites, the melt flow behavior during processing and the thermo-oxidative stability of the polymer improve significantly. Fearon *et al.* [17] attributed the positive effect of phenolic antioxidants to their interaction with peroxides. The trivalent phosphorous additives often help to improve the color of polymers [18–21].

3.2 Anti-block Agents

Anti-blocking agents prevent sheets of film from sticking together. Anti-block agents can be employed to prevent sheets of film from sticking together, while anti-static agents may be used to dissipate electrostatic charge buildup [22].

3.3 Antistatic Agent

Slip agents can be used to decrease the frictional properties of the plastic materials. Anti-static agents may be used to dissipate electrostatic charge buildup [22].

During processing and machining, static charge develops on the surface of polymers due to its non-conducting nature. Polymers attract dirt, cause fabric cling, and makes plastic film stick together. In electrical equipment, static charge may cause interference, and in certain environments such as grain elevators, for example, it can be an explosion hazard. Antistatic agents provide mechanism to the static charge by providing a hygroscopic surface layer that attracts atmospheric moisture, thereby increasing surface conduction [23].

3.4   Clarifying Agents

Clarifying agents reduce haze of the articles and are also used in processing. Plastics containing clarifying agents have higher crystallization initiation temperature during the cooling stage. They reduce the mold cooling time and hence cycle time reduction occurs. Clarifying agents have been successfully used both in polypropylene and linear low density polyethylene. The plastics with clarifying agents can be used in both blow molding and injection molding to affect results in the end products.

3.5   Slip Additives

Stearates, such as calcium and zinc stearates, are present in several commercial resins of both linear and long-chain branched polyethylenes. Stearates are not used as processing aids for linear low-density polyethylenes. They have strong evidence of promoting slip and aid in the reduction of instabilities with long chain branching material [24].

3.6   Processing Aids

Processing aids can significantly enhance product quality and processability, not only by eliminating surface defects, but also by reducing torque and power requirements as well as die plate pressure [25–26]. Processing aids eliminate flow instabilities or postpone them to higher flow rates. The end result is an increase of the productivity as well as an energy cost reduction, while high product quality is maintained.

Processing aids, particularly acrylic types, have been used in rigid PVC to increase fusion, decrease jetting and blushing down in blow molding, decrease parison draw down in blow molding, improve the rolling bank in calendering, improve thermoforming, and improve cell structure in foam extrusion [27].

In many instances it is difficult to process polyethylenes, especially where it comes to recycling. Studies were performed on the blending of different types of polyethylene in order to improve processing [28–32]. Typically, one would expect a processing agent (wax) to improve the melt flow of the polymer to be processed, without having a detrimental influence on the mechanical properties and thermal stability of that polymer.

In order to overcome these difficulties and to render the processes economically feasible, processing aids (PAs) are frequently used. PAs eliminate flow instabilities or postpone them to higher flow rates. The end result is an increase of the productivity as well as an energy cost reduction, while high product quality is maintained.

The processing aid polymer can significantly enhance product quality and processability, not only by eliminating surface defects, but also by reducing torque and power requirements as well as die plate pressure [26, 33].

3.7   Antifogging Agents

Antifogging agents prevent the condensation of moisture droplets inside film packaging that affect visibility of the plastic film. Non-ionic surfactants such as ethoxylated fatty acid esters are used. This wets the surface and runs the moisture into a continuous form. Antifogging agents, commonly used in food packaging, must meet the standards.

3.8   Antiblocking Agents

Antiblocking agents prevent surface sticking of the film layers together. Surface sticking is caused by static electricity and, to a lesser extent, by the interpenetration of molecules comprising the film which also is known as cold flow. It is a special type of release agent. Compounds such as fatty acids, waxes, other polymers such as polysiloxanes or fluorinated polymers will have limited solubility in the film or sheet and provide release properties.

Antiblocking agents are used in low concentrations primarily in PVC and other vinyl polymers.

Because polyethylene and polypropylene are in beads or granular forms, these additives are normally added in masterbatches. These masterbatches are added in proportion to add in lower concentration. Usually antiblocking agents are used in combination with calcium silicate or kaolin. Antiblocking agents do not dissipate static charge.

3.9   Heat Stabilizers

In processing, plastics can be subjected to heat or used to extend the life of the end products. Stabilizers are used to prevent degradation of material while processing. Heat stabilizers protect the material against the heat. During exposure to light by outdoor exposure, UV stabilizers are used to protect against deterioration in short and long term use.

The incorporation of UV stabilizer is necessary for protection against UV degradation. Enhanced UV light stability can affect the color stability and the mechanical properties of plastics. UV stabilizers protect PE from damage by sunlight in outdoor environments. Similarly, carbon black in PE mulch film protects and blocks the penetration of radiation.

Polymers, during their lifetime, are subjected to a number of degradation-initiating influences that result in the change, or even loss of their chemical and/or physical properties [34]. The need for chemical agents that would confer significant improvements on their properties has been evident since the very early stages of the polymer industry. These chemical agents are known as stabilizers and are incorporated into the polymer matrix at relatively low concentrations to combat degradation processes.

Heat stabilizers are targeted exclusively at PVC. Plasticizers are almost as dependent on PVC as heat stabilizers. Lead stabilizers have served the PVC industry as heat stabilizers. The European Commission has highlighted concerns about the dangerous nature of lead stabilizers which should be phased out, but did not indicate a specific timetable. The commission also highlighted that lead stabilizer would not be approved for use in PVC pipes after 2003. In America, unlike in Europe, window and door profiles were stabilized with tin stabilizers. However, in Asia, lead stabilizers still are in use along with PVC.

The wire and cable industry is changing its stabilizers to Ca/Zn. Organotin stabilizers are purely synthetic products. Organotin stabilizers are the most economically feasible material to use. The production of organotin stabilizer has the advantages of cost, availability, effectiveness in stabilization, and post processing characterization such as protection against weathering.

Stabilizers of fairly low thermal stabilizing performance include combinations of zinc, cadmium, lead, potassium, and barium which give better thermal stability.

3.10 Lubricants

Lubricants play an important role in plastics processing. Lubricants are outstanding and separating agents during processing. They have ability to gel and are water repellent. Moreover, they show some stabilizing effects.

Metal soaps are found as lubricant in numerous applications. It prevents excessive friction during processing so that high output together with protection from overheating of plastic materials is possible. Metal soaps provide excellent flow characteristics. They help plastics to be stressed dynamically in an extrusion or injection molding machine. The metal soaps have very little or no effect on the end products.

Metal soaps are used in polymers such as nylon, polycarbonate, polyester, polyethylene, polypropylene, and polystyrene as lubricating and separating agents. The high quality demanded led to the development of thermo-stable metal soaps to prevent discoloration at relatively high processing temperatures [35].

3.11   Plasticizers

Plasticizers are liquid, neutral substances. Particularly, the salvation and swelling power of plasticizers are used in plastics during processing.

Compounded with PVC, plasticizers change the rigid material so it is pliable and elastic across a range of temperature. With emulsion or micro-suspension PVC, the plasticizers form plastisol. The material with plasticizers in turn renders the possibility of room temperature shaping of end products produced by processes such as coating, dipping, and rotational casting.

Marked plasticizing effects have been seen with chloroparaffins of aliphatic chain length and degree of chlorination between 40–56%. Chloroparaffins contribute flame retardancy to compounds with subsequent cost reduction.

Plasticizers affect mechanical properties of the polymer [36]. They qualify as processing modifiers and lower polymer melt viscosity, even with addition of small amounts, and also migrate depending upon the molecular weight. Even though plasticizers are volatile in nature, they are very compatible. Usually additives are used in relatively small amounts. More addition of additive will cause undesirable effects in clarity due to accumulation on the surface of the film.

3.12   Coupling Agents or Surface Modifiers

Coupling agents improve the adhesion between polymer and filler, particularly inorganic. Providing a chemical bridge, coupling agents combine two incompatible materials. Silanes and titanates are the two commonly used coupling agents. They are commonly used in wood – plastic composition. Coupling agents are also used in polyethylene, polystyrene, poly(vinylchloride), and nylon.

3.13   Release Agents

In rotational molding, release agent is necessary to aid part removal. Release agents are either polymeric or oligomeric. Polysiloxanes are used widely as mold release agents for polyolefins, polystyrene, polyacetal, nylon, and ABS.copolymers.

Release agents reduce the coefficient of friction between polymer and surface and prevent sticking. They play a vital role in preventing products sticking during processing. When used in extrusion, they are called slip agents.

3.14   Flame Retardants

The amount of additives reduction maintains the mechanical standards but negatively affects the flame resistance of the materials [37]. In a flame retardant compound, addition of coupling agents or surface modifiers reduces its flame retardancy.

The bis(diphosphate) ester of resorcinol (RDP) provided an outstanding and improved flame retardant and mechanical properties to high performance thermoplastic materials such as PC, PPO, PC/ABS, PPO/HIPS and polyesters [38].

3.15   Pigments

Of all the pigments used in plastics, carbon black is unique. Besides acting as a colorant, it confers several additional properties. In the finer "particle" range it has a high ultraviolet light adsorbtion capacity, and thus protects the plastic from degradation on aging. This is extremely beneficial in cables exposed to the weather and sunshine. It may act as filler or as a reinforcing agent, whilst certain grades confer anti-static or conductive properties to plastic compounds [39].

Color reflects the natural energy. Color concentrates can aid safe handling. It is added directly into the feed. Color concentrates give optimal pigment dispersion and help to wet, minimize the wastage, and provide cost saving.

3.16   Light Stabilizers

Many polymer light stabilizers along with their relevant main functions have additional capacity to be heat stabilizers or antioxidants. These molecules often simultaneously bear fragments responsible for polymer protection against the destructive effects of both light and thermal oxidations. For instance, a number of hindered amine stabilizers (HAS) serve as both light and heat stabilizers [40–51].

3.17   Impact Modifiers

A polymer can be classified as brittle or ductile. Brittle polymers are characterized by having weak crack initiation and usually fail by crazing phenomena. Conversely, ductile polymers have significant crack initiation energy, significant crack propagation energy, and break through yielding phenomena. In order to improve the impact strength of a brittle polymer, elastomeric particles of adequate particle size and adhesion characteristics (in relation to the polymeric matrix) are usually included in the composite formulation. It is well established that rubber particles with low moduli act as stress concentrators in both thermoplastic materials favoring the dissipation of the impact energy by enhancing shear yielding and/or crazing, depending on the nature of the matrix, and, thus, improving the impact toughness of the blend. Also, the voids created by the cavitated rubber particles act further as stress concentrators [52].

The inclusion of rubber in polymers does, nevertheless, reduce the elastic modulus and the yield stress [53]. The phase separation between the polymer and the rubber is an important requirement, and mechanical resistance increases if the rubber has low elastic modulus in relation to the matrix, good adhesion to the matrix, adequate crosslinking, optimized average particle size and distribution, and low glass transition temperature [54]. The separating distance between the elastomeric zones also plays an important role in the toughening mechanisms.

In the case of PMMA and PS toughening, the impact modifiers (IMs) include: poly(vinyl acetate) (PVAc), copolymers of methacrylonitrile and ethyl acrylate, the ethylenevinyl acetate-vinyl chloride copolymer (EVAc-VC), the methyl methacrylate-methyl acrylate copolymer (MMMA) and the styrene-acrylonitrile copolymer (SAN). In blends of polymers based on styrene, the styrene-butadiene-styrene tri-block polymer (SBS) or the styrene-ethylene/butylene-styrene tri-block polymer (SEBS) are typical impact modifiers. Several elastomeric modifiers have been specifically synthesized for PVC. These are generally core-shell type rubbers with a rubbery core (typically poly(butadiene), styrene-butadiene rubber, poly(n-butyl acrylate) that has a rigid grafted outer shell (typically based on systems such as styrene-co-acrylonitrile (SAN), styrene-co-methyl methacrylate (S-MMA) copolymers, or PMMA. The PP matrixes contain EPDM (an elastomeric terpolymer from ethylene, propylene, and a nonconjugated diene) as the usual impact modifier. In order to improve its impact strength, namely the low temperature notched Izod impact, polycarbonates have been blended with a variety of low Tg, elastomeric impact modifiers, in particular core-shell rubbers such as PMMA-g-polybutadiene, PMMA-g-SBR, and PMMA-g-n-butyl acrylate (acrylic core-shell). Commercial impact modified PBT grades generally contain core-shell rubber modifiers, such as: PMMA-g-SBR, PMMA-g-poly(n- BuA) (acrylic core-shell rubbers), SAN-g-PBD and SAN-gpoly(n-BuA).

3.18 Blowing Agents

The use of physical and chemical blowing agents in producing foam plastics has been practiced for several years. Under usual processing conditions, the physical and chemical blowing agents influence the rheological properties of the compounded plastics. Extrusion of material containing blowing agent is an important industrial processing operation [55].

Future improvements in these processes may well depend on a better understanding of the processes and mechanisms involved. Studies of bubble morphology development and apparent rheological properties of thermoplastics in foam extrusion processes and the effect of processing variables on the quality of foam produced have been reported [56]. An important aspect of foam extrusion is the mechanics of bubble growth in the polymer matrix. Numerical and experimental studies of bubble growth during the microcellular foaming process and the influence of temperature, saturation pressure, molecular weight, and the nature of physical blowing agent have been reported recently [57]. Shear effects on thermoplastic foam nucleation have been studied on a twin screw foam extruder [58].

Several rheological aspects of thermoplastic foam extrusion have been described as a phenomenological model of the flow in an extrusion die, growth and physical properties of thermoplastic material with respect to extrusion foaming.

3.19 Nucleating Agents

Nucleating agents show heterogeneous nucleation effect. Increasing the nucleating agent has an effect on both high and low levels of the crystalline structure. Addition of nucleating agents improves crystalline temperature. Nucleating agents are widely used in commercial semi-crystalline polymers to improve the materials properties and to reduce the cycle times in injection molding. The effect of nucleation differs when using different nucleating agents. Different kinds of nucleating agent affect the mechanical properties of polypropylene materials by changing the crystalline structure. Polypropylene usually has higher tensile strength, high heat transformation temperature, and high hardness and transparence. The increase in crystallinity and the rate of crystalline is accelerated with nucleating agent effect. Nucleating agents influence the crystalline morphology of polypropylene materials [59].

3.20 Biocides

Biocides are used to protect polymers from the influence of microorganisms. There has been focus on the efficiency of the application for various classes of polymers. Numerous investigations have focused upon the efficiency for various classes of polymers [60]. Action of biocide is semi-quantitative [61]. As bio-overgrowth of the materials increases with time, kinetic methods of investigation may be suitable for determination of the inhibiting cation of biocides [62–63].

3.21 Fillers

Calcium carbonate is one of the most widely used fillers and the cheapest commercially available inorganic material. Being inexpensive, it can be used at high loading to reduce the cost of the production [64–73].

Polymers are often mixed with various particulate additives and fillers in order to produce a new class of materials termed polymeric composites [74]. This combination of materials brings about new desirable properties. For example, mineral fillers are added into the polymer matrix to improve mechanical properties, dimensional stability, and surface hardness. The effect of fillers on mechanical and other properties of the composites depends strongly on their shape, size and size distribution of the primary particles and their aggregates, surface characteristics, and degree of dispersion and distribution [75].

Of the various mineral fillers used, calcium carbonate ($CaCO_3$) is one of the most common, due mainly to its availability in readily usable form and low cost [76]. However, the incompatibility of its high energetic hydrophilic surface with the low-energy surface of hydrophobic polymers, e.g., polyethylene (PE) and polypropylene (PP), is a particular problem. For this and other reasons, the surface of calcite is often rendered organophilic by a variety of surface modifiers such as silanes, titanates, phosphates, and stearic acid.

Particulate, inorganic additives (fillers) are commonly added to commercial thermoplastic and thermosetting resins to achieve economy, as well as to modify favorably certain properties, such as stiffness, heat distortion, and molding. However, usually there is a trade-off, involved with certain important properties, such as toughness and ultimate elongation, usually deteriorating. In fact, one of the major drawbacks of the use of mineral filler is wear on the processing equipment. There are several reports in the literature concerning the influence of the inclusion of $CaCO_3$, for example, as mineral filler in thermoplastic matrices [76–79].The presence of the fillers affects to a great extent the morphology and the structure of semicrystalline polymers. Perhaps the most striking and important changes are observed in the mechanical behavior of composites [80–81]. The reason is that fillers drastically affect the morphological structure of the polymer, especially when it undergoes a phase transition, such as crystallization, after the introduction of fillers. Depending of the filler nature and its concentration, the crystallinity changes as well as the topology of the non-crystalline region [79, 82].

The commonly used polymer materials are usually manufactured not only by mixing several different macromolecules, but also by incorporating certain solid "filler" particles into the materials, in order to improve the modulus, impact strength, appearance, conductivity, or flammability of the materials [83].

The introduction of mineral fillers into a plastic can improve some mechanical properties, but it affects some other properties like impact strength. Very fine particles can minimize the negative effects [84]. Mineral fillers modify the mechanical properties by their nature, size, shape, and distribution.

Fillers with respect to plastics are inert substances. They are used to increase the characteristic strength in plastics. Normally fillers are used to lower the cost of the product. Use of filler beyond certain limits leads to decrease in mechanical properties. The selection of filler is based on the final properties of the end product and the plastics used. In order to know the effect of concentration of filler, it must be selected according to size by screening and chemical compatibility with the plastics [85–86].

Calcium carbonate is a commonly used filler in plastics processing. It improves the ductile performance of LLDPE and HDPE films. Coated or treated calcium carbonate helps to wet the plastic material to the mineral surface and improve the processing. The coating in calcium carbonate also helps to improve greatly the dispersion of the mineral into the polymer matrix [87]. High filler loadings, however, may adversely affect processability, ductility, and strength of the compound.

For filled polymers, it has been observed experimentally [88–90] and numerically [91] that the degree of strain hardening generally decreases with increasing particle concentration. The chemical nature of filler is one of the fundamental characteristics that determine its reactivity.

Inorganic filler is often added into PE to form a composite, in which the filler serves to improve the mechanical properties and lessen the costs as well. Since the natural photo-oxidation of PE causes the deterioration of the mechanical properties, and even fracture when a PE product is exposed to outdoor application, it is necessary to investigate the weathering stabilities of PE composites [92].

The addition of fillers such as carbon and ceramics (silica, alumina, aluminum nitride, etc.) is commonly used to induce thermal conductivity into conventional polymers. The higher thermal conductivity can be achieved by the addition of high volume fractions of suitable use of a filler. Fillers have to form a random close packed structure to maximize a pathway for heat conduction through the polymer–matrix [93]. Mineral fillers are specially selected to give the compounds particular characteristics and improve the price-performance ratio.

Conducting fillers are used in applications such as carpeting, linoleum, belting, and footwear, including metal filament and carbon black. Conducting fillers provide a conducting path through the polymer. Quaternary ammonium compounds, textile fibers, or polyelectrolytes act as external agents. Poly(ethyleneglycol) of higher molecular weight, fatty acid esters, long chain sulfonates, and phosphates are used as internal agents. However, poly(ethyleneglycol) dichloride is a typical polyelectrolyte used as an external antistatic agent.

3.22   Fundamentals

- Additives influence the physical properties and resistance to deterioration in plastics
- Complex structure of plastics and additives dictates the physical properties of the material
- Properties of plastics can be modified through the use of chemical additives
- Plasticizers are used to adjust the flexibility without deterioration of physical properties
- Solid fillers are added to reduce the cost or modify mechanical, thermal, or electrical properties
- Processing aid lowers processing temperature and hence increases the output
- Processing aid reduces defects and rejects.

References

1. Klender, G.J., Glass, R.D., Kolodchin. W., and Schell. R.A., paper presented at Annual Technical Conference of Society of Plastics Engineers, Washington D.C., April 29–May 2, 1985, pp. 989–996.
2. Kletecka, G., *Polyolefins VII International Conference*, February 24–27, 1991, pp. 254–273.
3. Asay, R.E., *Polyolefins V International Conference*, February, 1987 pp. 210–223.

4. Binsbergen, F., Heterogeneous nucleation in the crystallization of polyolefins: part 1. Chemical and physical nature of nucleating agents. *Polymer* 1970;11:253–67.
5. Bruckner, S., Meille, S.V., Polymorphism in isotactic polypropylene. *Progress in Polymer Science* 1991;16:361–404.
6. Gaylord, N.G., Mehta, R., *J Polym Sci: Polym Lett* 1982, 20, 481.
7. Hogt, A., SPE *ANTEC Tech Papers*, 1988, 34, 1474.
8. Callais, P.A. Kazmierczak, R.T. SPE *ANTEC Tech Papers* 1989, 35, 1368.
9. Callais, P.A. Kazmierczak, R.T. SPE *ANTEC Tech Papers*, 1990, 36, 1921.
10. Ganzeveld, K.J., Janssen, L.P.B.M. *Polym Eng Sci*, 1992, 32, 467.
11. Sache, A.N., Rao, G.S.S., Devi, S. *J Appl Polym Sci* 1994, 53, 239.
12. Kim, B.J., White, J.L. *Int Polym Process* 1995, 10, 213.
13. Samay, G., Nagy, T., White, J.L. *J Appl Polym Sci* 1995, 56, 1423.
14. Weihua, L., Jingyuan, W., Yaoxian, L., Yuwei, L., Xinyi, T. *J Polym Eng* 1995 to 6, 15, 271.
15. Heinen, W., Duin, M. SPE *ANTEC Tech Papers* 1997, 43, 2017.
16. Cha, J., White, J.L. *Polym Eng Sci* 2001, 41.
17. Fearon, P.K, Phease, T.L, Billingham, N.C, Bigger, S.W., A new approach to quantitatively assessing the effects of polymer additives. *Polymer* 2002;43:4611–8.
18. Klemchuk, P.P., Horng, P.L., Perspectives on the stabilization of hydrocarbon polymers against thermo-oxidative degradation. *Polym Degrad Stab* 1984;7:131–51.
19. Drake, W.O., Cooper, K.D., Recent advances in processing stabilization of polyolefins. Proceedings of SPE polyolefins VIII. Houston, Texas: *International Conference Society of Petroleum Engineers*; 1993. p. 414–27.
20. Zweifel, H., *Stabilization of polymeric materials*. Berlin: Springer; 1998.
21. Drake, W.O., Pauquet, J.R., Todesco, R.V., Zweifel, H., Processing stabilization of polyolefins, *Angew Makromol Chem* 1990;176/177:215–30.
22. Joyner, R.S. (1971). Polyethylene, *Modern Plastics*, 48, pp. 72, 77, and 80.
23. Johnson, K., *Antistatic Agents Technology and Applications*; Noyes: Park Ridge, NJ, 1972.
24. Hatzikiriakos, S.G., Kazatchkov, I.B., and Vlassopoulos, D., *J. RheoL*, 41(6), 1299 (1997).
25. Rudin, A., Blacklock, J.E., Nam, S., and Worm, A.T., ANTEC 1986, Soc. Plastics Engrs., *Tech. Papers*, 32,1154–1158 (1986).
26. Athey, R.J., Thamm, R.C., Souftle, R.D., and Chapman, G.R, ANTEC 1986 Soc. Plastics Eng'rs. *Tech. Papers*, S2, 1149–1152 (1986).
27. Casey, William, J., i Okano, Kenji., *Journal of Vinyl Technology*, Volume 8, Issue 1 , Pages 37–40; 2004.
28. Liang, J.-Z., Ness, J.N., *Polym. Test* 16 (2) (1997) 173.
29. Perez, R., Rojo, E., Fernandez, M., Leal, V., Lafuente, P., Santamaria, A., *Polymer* 46 (19) (2005) 8045.
30. Hussein, I.A., Hameed, T., Abu Sharkh, B.F., Mezghani, K., *Polymer* 44 (16) (2003) 4665.

31. Liang, J.Z., *Polym. Test* 21 (1) (2002) 69.
32. Wong, A.C.Y., *J. Mater. Process. Technol.* 48 (1–4) (1995) 627.
33. Wigotsky,V., Plastics Engineering, 43, No.2, Feb.1987, p. 21–6.
34. Habicher, W.D., Bauer, I., in: Halim, H.S. , (Ed.), *Handbook of Polymer Degradation,* Marcel Dekker, New York, 2000, p. 81.
35. *Industrial Lubrication and Tribology,* Vol 43 No 1. 1991, pp. 3–4.
36. Stevens, M.P., *J. Chem. Educ., 1993, 70 (6), p. 444.*
37. Petrucci, L.J.T., Monteiro, S.N., Rodriguez, R.J.S., and d'Almeida, J.R.M., Low-Cost Processing of Plastic Waste Composites, *Polymer-Plastics Technology and Engineering,* 45: 865–869, 2006.
38. Durairaj, Raj B.; Jesionowski, Gary A., Proceedings of the Conference on *Recent Advances in Flame Retardancy of Polymeric Materials* (2005), 16 93–103.
39. *Pigment and Resin technology,* JANUARY 1972, p. 25.
40. Zeynalov, E.B., Allen, N.S., Effect of micron and nano-grade titanium dioxides on the efficiency of hindered piperidine stabilizers in a model oxidative reaction. *Polym Degrad Stab* 2006;91(4):931–9.
41. Zaharescu T, Kaci M, Hebal G, Setnescu R, Setnescu T, Khima R, et al. Thermal stability of gamma irradiated low density polyethylene films containing hindered amine stabilizers. *Macromol Mater Eng* 2004; 289(6):524–30.
42. Gijsman P, Gitton-Chevalier M. Aliphatic amines for use as long-term heat stabilizers for polypropylene. *Polym Degrad Stab* 2003;81(3): 483–9.
43. Schroeder, H.F., Zeynalov, E.B., Bahr, H., "Analysing the content of antioxidants in PP materials". *Polym Compos* 2002;10(1):73–82.
44. Schwetlick K, Habicher WD. Antioxidant action mechanisms of hindered amine stabilisers. *Polym Degrad Stab* 2002;78(1):35–40.
45. Allen, N.S., Ortiz, R.A., Anderson, G.L., Sideridou, I., Malamidou, E., Comparison of the thermal and light stabilizing action of novel imine and piperazine based hindered piperidine stabilizers in polyolefins. *Polym Degrad Stab* 1994;46(1):85–91.
46. Gugumus, F., Mechanisms of thermooxidative stabilisation with HAS. *Polym Degrad Stab* 1994;44(3):299–322.
47. Yongcheng, Y., Thermal oxidation of polypropylene containing hindered piperidine compounds. *Polym Degrad Stab* 1992;37(1):11–7.
48. Gijsman, P.,Gitton, M., Hindered amine stabilisers as long-term heat stabilisers for polypropylene. *Polym Degrad Stab* 1999;66(3):365–71.
49. End, M.J., Davis, L.H., Vulic, I., In: *Proceedings of the second world congress e polypropylene in textiles,* Queensgate Huddersfield, UK; 2000. p. 295–313.
50. Gensler, R., Plummer, C.J.G., Kausch, H.H., Kramer, E., Pauquet, J.R., Zweifel, H., Thermo-oxidative degradation of isotactic polypropylene at high temperatures: phenolic antioxidants versus HAS. *Polym Degrad Stab* 2000;67(2):195–208.

51. Gugumus, F., New trends in the stabilization of polyolefin fibers. *Polym Degrad Stab* 1994;44(3):273–97.
52. Walker, Collyer, A.A., in: Collyer, A.A. (Ed.), *Rubber Toughened Engineering Plastics*, Chapman & Hall, London, 1994, p. 29.
53. McGrath, G.C. in: Collyer, A.A. (Ed.), *Rubber Toughened Engineering Plastics*, Chapman & Hall, London, 1994, p. 57.
54. Bigg, D.M., Preston, J.R. and Banner, D., *Polym. Eng. Sci.*, 1976, 16, 706; Gonzalez, H., *J. Cell. Plast.*, 1976, 12, 49–58.
55. Ham, C.D. and Villamizar, C.A., *Polym. Eng. Sci.*, 1978, 18, 687.
56. Ramesh, N.S., Rasmussen, D.H., and Campbell, G.A., *Polym. Eng. Sci.*, 1991, 31, 1657.
57. Lee, S.T., *Polym. Eng. Sci.*, 1993, 33, 418.
58. Wang, K,F., Mai, K.C., *Syn Resin Plas* 1999;16(6):49.
59. Popov, A.A., Rapoport, N.Ya. & Zaikov, G.E., *Oxidation of Stressed Polymers*. Gordon and Breach, Philadelphia, PA, 1991.
60. Andrejuk, E.I., Bilai, W.I., Koval, E.Z. & Kozlova, I.A., *Microbial Corrosion and Their Stimuli*. Naukova Dumka, Kiev, 1980.
61. Kharitonov, V.V., Psikha, B.L. & Zaikov, G.E., *J.Polymeric Mater.*, 26 (1994) 121.
62. Mura, C., Yarwood, J., Swart, R., Hodge, D., *Polymer*, 41, 24, 8659–8671, 2000.
63. Burley, J.W., Clifford, P.D., *J. Vinyl Additive Technol.*, 10, 2, 95–98, 2004.
64. Bartosiewicz, L., Kelly, C.G., *Adv. Polym. Technol.* 7 (1987) 21–28.
65. Menczel, J., Varga, J., Influence of nucleating agent on crystallization of polypropylene, *Therm. Anal.* 28 (1983) 161–165.
66. Fujiyama, M., Wakino, T., Role of filler particles on nucleation of polypropylene, *J. Appl. Polym. Sci.* 42 (1991) 2739–2742.
67. Khareh, A., Mitra, A.A., Radhakrishnan, S., Effect of $CaCO_3$ on the crystallization behavior of PP, *J. Mater. Sci.* 31 (1996) 5691–5702.
68. Pukanszky, B., Polypropylene, in: Karger–Kocsis, J. (Ed.), *Structure Blends and Composites*, vol. 3, Chapman & Hall, Britain, 1995.
69. Pukanszky, B., Vanes, M., Maurer, H.J., Vörös, G.Y., Micromechanical deformations in particulate filled thermoplastics: volume strain measurements, *J. Mater. Sci.* 29 (1994) 2350–2355.
70. Pukanszky, B., Vörös, G.Y., Stress distribution around inclusions and mechanical properties of particulate-filled composites, *Polym. Comp.* 17 (1996) 384.
71. Farris, R.J., Matrix/filler debonding in polymer composites, *Rheo. J.* 12 (1998) 315–321.
72. Jancar, J., Dibenedetto, A., Mechanical properties of ternary composites of PP with inorganic fillers and elastomer inclusions, *J. Mat. Sci.* 30 (1995) 1601–1608.
73. Karger–Kocsis, J., *Polypropylene, Structure, Blends and Composites*, first edition, Chapman & Hall, Britian, 1995.

74. Osman, M.A., Suter, U.W., *Chem. Mater.* 14 (2002) 4408.
75. Chan, C.M., Wu, J., Li, J.X., Cheung, Y.K., *Polymer* 43 (2002) 2981.
76. Thio, Y.S., Argon, A.S., Cohen, R.E., Weinberg, M., *Polymer* 43 (2002) 3661.
77. Chacko, V.P., Farris, R.J., Karasz, F.E., *J Appl Polym Sci* 1983;28:2701.
78. Qiang. F., Guiheng,W., *Polym Eng Sci* 1992;32:94.
79. Chacko VP, Karasz FE, Farris RJ, Thomas EL. *J Polym Sci Phys* 1982;20:2177.
80. Jiang, H., and Pascal Kamdem. D., Journal of vinyl & additive technology, 2004, Vol. 10, No. 2.
81. Qiang F, Guiheng W. *Polym Eng Sci* 1992;32:94.
82. Bozveliev LG, Kosfeld R, Uhlenbroich Th. *J Appl Polym Sci* 1991;43:1171.
83. Meijer HEH, Lemstra PM, Elemans PHM. Macromol Chem, Macromol Symp 1988;16:113.
84. Riley, A.M., Paynter, C.D., McGenity, P.M. and Adams, J.M., Factors affecting the impact properties of mineral filled polypropylene. Plust. Ruhh. Process. Applit., 1990, 14(2), 85–93.
85. Levenhagen, A.W., *Sixth Annual Technical Session*, Sect. 4, p. 1, Reinforced Plastics Division, Society Plastics Industry, New York, N. Y., 1951.
86. Linzmever, L.G, *Seventh Annual Technical Session*. Sect. 4. p. 1, Reinforced Plastics Division, Society Plastics Industry; New York, N. Y., 1952.
87. Ruiz, F. A., High performance mineral reinforcement concrete for LLDPE and HMW-HDPE blown film extrusion, *Journal of Plastic Film and Sheeting* 2002; 18; 25–31.
88. Kobayashi, M., Takahashi, T., Takimoto, J., Koyama, K., Flow-induced whisker orientation and viscosity for molten composite systems in a uniaxial elongational flow-field, *Polymer* 36 (1995) 3927.
89. Kobayashi, M., Takahashi, T., Takimoto, J., Koyama, K., Influence of glass beads on the elongational viscosity of polyethylene with anomalous strain rate dependence of the strain-hardening, *Polymer* 37 (1996) 3745.
90. Le Meins, J.F., Moldenaers, P., Mewis, J., Suspension of monodisperse spheres in polymer melts: particle size effects in extensional flow, *Rheol. Acta* 42 (2003) 184.
91. D'Avino, G., Maffettone, P.L., Hulsen, M.A., Peters, G.M.W., Numerical simulation of planar elongational flow of concentrated rigid particle suspensions in a viscoelastic fluid, *J. Non-Newtonian Fluid Mech.* 150 (2008) 65.
92. Yang, R.,Yu, J., Liu, Y., and Wang, K., *Polymer Degradation and Stability* 88 (2005) 333–340.
93. Lee, G.W., Park, M., Kim, J., Lee, J.I., Yoon, H.G., Enhanced thermal conductivity of polymer composites filled with hybrid filler. *Compos A* 2006;37:727–34.

4

Plastics Processing

Polymer processing emerged as a separate, identifiable engineering discipline in the late 1950s and 1960s. The reasons for this exciting development can be attributed to important advances that took place in the late 1980s and 1990s. Plastics processing faces challenges and opportunity to develop techniques and needs with materials in terms of intellectuality and modernization. Some of the important developments are [1].

- Improvement in polymer properties to suit different applications
- Progress in engineering plastics
- Scientific compounding to create copolymers, block copolymers, graft copolymers, etc.
- Computer revolution in plastics manufacturing and processing technology
- Instruments to measure the polymer properties.

Plastics processing provides different methods to produce different products with the development of thermoplastic materials. Thus, focus on plastics processing is characterized with technologies, knowledge, and hi-tech involvement [2].

In the thermoplastic materials, production and processing occur in the molten state. Knowledge of flow behavior of the material is essential for all forms of production and processing [3]. Also, polymer processing is a multidisciplinary field that fuses with polymer physics and polymer chemistry.

Plastic materials are the most versatile material and contribute in several aspects and in their initial stage as powder or granules. Plastic processing techniques are used to utilize the powder or granules for finished parts. The ability of thermoplastics to melt and re-harden has been exploited in many different processing methods [3–5].

Temperature and pressure are known to be two of the most influential variables in the rheological characteristics of polymer melt in plastics processing. It has long been realized by those in practice that the flow behavior of molten polymer is highly dependent upon these two variables.

4.1   Focus on Plastics Processing

Plastics processing can shape material and improve its properties in terms of end products produced. In plastics processing, material undergoes a common practice of mixing, melting, plasticizing, shaping, and finishing. Thermoplastic materials have sufficient melt strength to remain integral during processing.

Plastics are available in the form of pellets, granules, and powder which can be extruded, blow molded, injection molded or rotational molded, to finished products. Knowledge of plastics to production to their end use properties has become important in view of material development [6–8].

Major processing methods like extrusion, injection molding, thermoforming, and rotational molding have been subjected to significant technological progress [9]. The choice of processing method has both technical and economical aspects between material and end product properties [10, 11].

Plastics processing requires external heating to melt and the resulting frictional heat impacts finished product quality. The plastic melt is highly viscous during processing. The polymer melt depends on molecular weight, molecular weight distribution, chain branching, shear rate, and shear stress [12].

Plastics shrink in molds during the cooling process. Shrinkage for amorphous plastics is less than for semi-crystalline plastics. Shrinkage is larger in the thickness direction than other directions. Orientation of plastic occurs in the direction of flow.

4.2   Injection Molding

The injection molding machine is similar to an extruder, except the die is replaced by a valve nozzle and the screw can also move axially to act as a piston to pump the melt. Injection molding requires a polymer with a low molecular weight to maintain low viscosity during the injection process. In contrast, extrusion requires polymer with higher molecular weight for better melt strength. The extrusion process produces continuous linear profiles by forcing a melted thermoplastic through a permanent mold which is called a die, where it takes shape and cools, whereas in injection molding, the mold is fixed and the product will be discharged after the injection and cooling. Extrusion involves heating, shaping, and cooling the material continuously.

The injection molding process produces three-dimensional items, even though both processes also use screws to convey, pump, and blend the heterogeneous components [13]. In injection molding and extrusion, screw design also plays a role during processing.

Injection molding has a rapid production rate on complex shape products with good dimensional accuracy and surface finishing. Optimized processing condition helps to achieve excellent product quality. Plastic melt in injection molding and extrusion tends to align in the direction flow. This produces markedly greater strength in the direction of orientation.

4.2.1   Injection Molding – Machine

4.2.1.1   Ram Injection Molding Machine

Earlier ram injection molding machines were used for manufacturing molded parts. This had disadvantages of poor plastication and inadequate mixing and hence thermal gradients in melt cause inconsistent flow behavior. Due to variable injection pressure, ram compresses the mixture to vary the material from solid granules to melt causing significant pressure differences and cycle varies. High pressure drops occurred with a ram injection molding machine due to the presence of torpedo. Because it is difficult to measure shot size, any variation of the bulk density of the feed affected the shot weight.

4.2.1.2   Screw Injection Molding Machine

The screw rotates and axially reciprocates by a hydraulic motor which results in melting, mixing, and pumping of the polymer. Screw injection provides fast plasticization and also decreased cycle time. The polymer melt move forward for injection is allowed by the hydraulic system. The hydraulic system controls the axial reciprocation of the screw. A valve prevents back flow of the polymer from the mold cavity. The screw acts as a plunger to inject the polymer melt in the mold. Precise timing of all pressures and temperatures is an especially critical factor for this high-speed process. Opening and closing the mold is achieved hydraulically. A system of mold locking, temperature control, guide pins, and ejector pins is also necessary.

4.2.2   Injection Unit

Injection unit consists of barrel, screw, and nozzle. The flow starts from machine barrel through the nozzle of the injection unit into the mold cavity. The injection unit melts the polymer resin and injects the polymer in the mold. Before the packing pressure is released, the molten plastic freezes at the gate. No material can be further packed when the gate, a narrow entrance connecting the injection unit and mold, freezes. Figure 1 illustrates the schematic diagram of an injection molding machine.

4.2.2.1 Barrel

The barrel is made of a steel cylinder that contains a reciprocating screw. Heating media, usually electrical heaters, are fitted around the barrel as the heat source. Temperatures are measured by thermocouples installed within the different zones of the barrel. Barrel temperature should be controlled within a small variation of the set-point. The barrel has three zones,

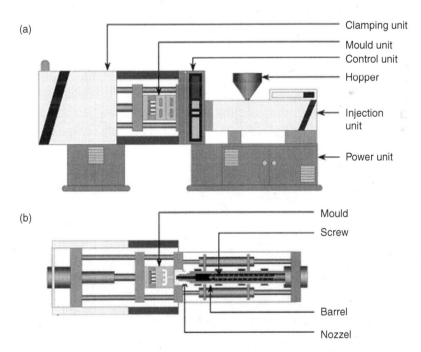

Figure 1. Schematic diagram of an injection molding machine (a) overview (b) section (Reproduced with permission from Dumitrescu, O.R. *et al. Polymer Testing* 24 (2005) 367–375. © 2004 Elsevier Ltd. All rights reserved.) Ref.[14].

namely feed, compression, and metering zones, excluding the nozzle. The barrel of the injection molding machine is heated by the electric heater bands.

The barrel of the injection molding machine supports the reciprocating plasticizing screw. It is heated by the electric heater bands. Increasing heat profile across the barrel may not be the most appropriate profile for all parts. It depends on the size of product and mold; the heat profile may be a reverse, hump, or flat profile. Some adjustments may be necessary to obtain the optimum process conditions and part quality. Fixed temperature profile from hopper to nozzle is a constant profile to utilize the plasticizing capacity.

Generally, the barrel temperature of the injection molding machine should increase from the hopper to the nozzle gradually. The material enters into the barrel and is heated slightly higher than the melting points. Mixed profile is to lower temperature at the nozzle to prevent drooling or stringing. Rising temperature profile from hopper to nozzle is to allow moderate fusion of resin.

4.2.2.2 Screw

In the plasticization, the screw rotates, hence shears and melts the material in the barrel and conveys the melt to the screw front for the next shot.

The reciprocating screw is used to compress, melt, and convey the material (similar to the extruder screw). The difference is that:

- First, as the screw rotates it retracts to fill the shot chamber
- Then it pushes forward to feed the molten plastic through the nozzle into the mold.

A lower screw speed will improve dispersion of additives and will help to maintain low melt temperatures with thick parts molding.

Specific output has an opposite effect on the delay volume and screw revolution delay [15]. Higher output will result in a larger delay volume and a smaller screw revolution delay.

In the closed barrel system, the plastic is melted by mechanical energy from a rotating screw and heat transfer from the high temperature barrels. During shutdown, the distribution of melt material may be changed. Rotating screw speed is reduced to zero in a very short time. The residual solid polymer may be melted while the barrel is being quenched, especially when the screw channel is only partially full of polymer, or the depth of screw channel is large. The operation is time consuming.

Shutdown of the injection molding machine is undesirable, but occurs from time to time in any production setting. In such shutdown, molten polymer is trapped in the barrel of the molding machine, and thermal and oxidative degradation may accelerate and lead to severely degraded polymer and poor part properties. In short stoppages, it may not be an issue for some polymers such as polyethylene. The shutdowns are a concern for polymers containing residual unsaturation, like ABS [15].

4.2.2.3   Clamping Unit

The clamping unit of an injection molding machine provides the motion needed for mold closing and opening, and produces the forces that are necessary to clamp the mold. The clamping unit holds the mold together, opens and closes it automatically, and ejects the finished part. It has principal components including tie bars, stationary and movable platens, and a mechanism for mold opening, closing, and clamping.

The stationary platen carries a force from the four nuts at its four corner regions to balance the reaction force as shown in Figure 2.

4.2.2.4   Hydraulic Unit

A hydraulic unit is composed of hydraulic components with system pressure and is controlled by proportional valve. The

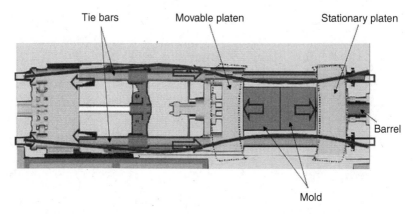

Figure 2. Schematic diagram of the tie bars, movable platen and stationary platen. (Reproduced with permission from. Sun, S.H. *Computers in Industry* 55 (2004) 147–158 © 2004 Elsevier B.V. All rights reserved.). Ref. [16].

hydraulic system is defined as pumps, motors, actuators, valves, piping, etc. used in conjunction with hydraulic fluid to provide useful energy. Pumps and motors provide high pressure with hydraulic fluid. The cycle starts with diversion of hydraulic fluids to clamping section, and the mold begins closed.

The hydraulic pressure or mechanical pressure is applied to make sure all of the cavities within the mold are filled. Plastics are allowed to cool within the mold. The mold is opened by separating the two halves of the mold. The product is ejected from the mold with ejecting pin. The runners and sprue are trimmed off and recycled.

4.2.3   Mold

Injection molds are typically complex tools. They are expected to be efficient and reliable in operation. The molds have to be cost and time effective at their design and manufacturing stages. Figure 3 illustrates the mold design and parts in a three dimensional view.

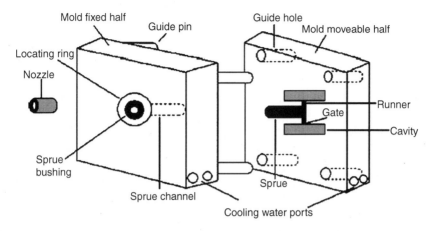

Figure 3. Typical Plastics injection mold (Reproduced with permission from Tatara, R.A. *et al. Journal of Materials Processing Technology* 176 (2006) 200–204 2001 © 2006 Elsevier B.V. All rights reserved.). Ref. [24].

Mold

- Consists of two halves
- Is brought together and clamped into position.
- Is kept at a constant temperature
- Is in the cool condition; hot molten plastic is then forced under pressure.

Every mold contains cooling channels to help to maintain uniform heat distribution throughout the tool. It may contain internal cooling channels with circulating fluid [17, 18]. It is held together under pressure by the tie bars that prevent leaking (bleeding) of the molten polymer through the parting line. The nozzle forms a seal between the injection system and the mold.

Molds that are subjected to many, short production runs are also in danger of premature wear. Check the mold for hot spots and try to achieve uniform cooling. Make sure both mold halves are at equal temperature. A temperature variation in one section of the mold surface also can create a non-uniform flow front.

The cost of tooling is the highest of all plastic production in the plastics processing sector [19]. Good surface finish is obtained by polishing, leading to more expensive molds. Economy in the mold making puts pressure on the use of not so smooth surface finish of the cores. The molded part is difficult to separate from the polished mold due to the local buildup of adhesion forces. To minimize these problems, the surface finishing is made in the direction of ejection.

In the mold design, design of the mold with specific supplementary geometry, usually on the core side, is quite complicated by the inclusion of projection and depression [20]. Important factors of designing the mold include mold size, number of cavity, cavity layouts, runner systems, gating systems, shrinkage, and ejection system [21–23].

Both mold and machine are compliant and deform upon loading can affect the part quality. The minimization of mold deflection is essential to tight tolerances in plastic parts during manufacturing. To determine the final product geometry, it is critical to understand mold deflection during injection molding [25].

Mold filling is considered with various factors of part and mold designs. Material selection and process setup have to be considered to ensure that the mold can be filled volumetrically.

The friction forces develop between the polymer surface and the surface of the mold, which is usually made of steel [26]. Also, it depends on resin and mold used, and the non-linear, distributed and time varying process dynamics. Manipulating the machine settings and process dynamics is necessary to control process variables [27].

In the injection molding process, mold leads to tribological (frictional) problems. Plastic materials while processing liberate gases, resulting in corrosion and wear of the mold material. In addition, the flow of molten plastic in contact with mold surfaces, problems of abrasion, adhesion, release, and fatigue commonly occur [28]. Appropriate surface modification technique will resolve the tribological problems. In injection molding process, the matching mold halves decide the dimension of the end product.

4.2.3.1 Gate

The gate location is to be changed so as to move the weld line to a non-stressed area away from load bearing areas. Processing can influence the strength and cosmetics of the weld line, but it cannot eliminate the root causes in the material or part or tool design. Optimize weld line performance by picking a gate location that will allow the polymer to continue to flow and merge after flow fronts recombine [29]. A multiple gate system reduces pressure. However it may cause visible weld lines and require complex runner systems.

The cross-section of the gate is smaller than that of the runner, so that the part can be separated from the runner. A larger gate reduces viscous heating, permits lower velocities, and allows the application of high packing pressure. Gate location is to be selected to fill the mold uniformly and the weld lines and air vents are positioned properly. Packing of the cavity ends when the gate freezes.

Gate design includes selection of type, dimensions, and location. Gate design is of great importance to part quality and productivity. Single gate is generally preferred, unless it is otherwise necessary to use multiple gates. Multiple gates create weld lines. The cross section of the gate is smaller than that of the runner and part so that the part can be separated from the runner. Gate location should be selected in such a way that uniform mold filling is ensured. It is dictated by the part and mold design, the part specs (appearance, tolerance, concentricity, etc.), the material, the fillers, and economic factors (tooling cost, cycle time, scrap volume, etc.). Weld lines can be eliminated by sequentially opening and closing valve gates. A multiple-gate system reduces pressure but may cause visible weld lines and require complex runner systems.

The gate connects the runner to the part. The physical dimensions of the gates are usually the smallest out of both the runner and the part. However, it should be noted that the gates cannot provide a good control of the polymer flow due to the heat transfer effects, and the pressure build up on the entry to the gate and its subsequent drop on the exit. The position of the gate is selected in such a way that the melt flow throughout the cavities is well balanced. The gates determine the flow field into the part cavity.

Gate design includes selection of type, dimension, and location. The material, filler, economic factors, mold design, and the part specifications dictate the gate location. It is of great importance to part quality and productivity. Single gate is generally preferred. In case of necessity, multiple gates are allowed.

When gates freeze, packing of the cavity ends. A larger gate reduces the viscosity heating, permits lower velocity and allows the application of high packing pressure. Gate freeze ends the packing of the cavity. The gates determine the flow field into the part cavity [23–30].

4.2.3.2 Runner

The runner system is one of the most influential factors on the injection molding process and the properties of the injected parts. They are full round, half round, and trapezoidal, and, in general, full round runner are preferable. Runners are to be recycled; hence they should be as short as possible and it is preferable to restrict the length and diameter to minimize the amount of material in the molding process.

The runner system is used to convey the plastic melt from the sprue to the gate. It is important in producing identical quality of plastic parts. Cost reduction can be carried out by using multi-cavity mold with a balanced runner system and the mold fills in the cavity at the same time. The runners deliver the plastic into the part cavities (usually multiple).

4.2.3.3   Sprue

The size of the sprue must be larger than maximum wall thickness of the molded part. The sprue delivers the molten plastic into the mold. The dimensions of the sprue depend primarily on the dimension of the molding and especially its wall thickness. The sprue must not freeze before any other cross-section in order to permit sufficient transmission of holding pressure. The sprue must be molded easily and readily.

4.2.3.4   Cavity

Cavity is important to improve the performance of the molded products and obtain precision form. Cavity controls, indirectly, the injection molding process through adjusting correlative parameters during filling. Cavity filling depends on the design of the runner system. Cavity geometries or runner will induce different molecular orientations [30, 31]. The cavity should be well vented to allow the trapped gases and air to escape with all plastics. Unless properly vented, the gas can hold the flow fronts apart.

Part quality is strongly dependent on the processing conditions from plasticization to shaping and solidification of polymer inside the mold cavity [32, 33]. In injection molding, the characteristics of cavity pressure play a dominant role throughout all phases of the end product. The end product quality can be characterized by part weight, thickness, or other dimensional features like shrinkage and warpage [34–36]. In the temperature differences of molten resin and the mold, an uneven temperature field exists around the product cavity during cooling [37].

4.2.3.5   Nozzle

A nozzle will help the material to get pumped in the mold; there are different types such as shut-off nozzle, etc. Both open nozzles and needle shut-off nozzles can be used. Open nozzles are often preferred to shut off devices because of their streamlined design and because they are more readily replaced, particularly if a change-over is made to another color. However, solidified melt in the nozzle orifice can be removed more easily and completely from a shut-off nozzle.

Nozzle designs with positive shut-off devices have been successfully used. The gas must be free to escape through the nozzle. Material freeze off in the nozzle or malfunction of the positive shut-off device could develop pressure to cause blow-back of the material through the feed zone and hopper or create hazardous conditions. In such cases, conventional free flow and reverse taper type fitted with a heater band for temperature control of the nozzle prevents nozzle drool or freeze off of the material and is used and in nylon processing. Sprue cutter is associated with the nozzle to help the process. A nozzle forms a seal between the injection system and the mold.

4.2.3.6   Vent

Vents should be placed in other locations as well, including the runner system, weld line regions, and other areas of possible gas entrapment. Air vent design is important because its function is to release the air inside the cavity when the mold is closed. Short shot will happen if air is trapped inside the mold.

4.2.3.7   Ejection System

The frictional forces develop during ejection between the polymer surface and the surface of the mold [38]. The ejection system requires the knowledge of involved forces. The ejection system assumes a relevant importance in the product quality, where the parts are difficult to extract from the mold cores [39–41].

4.2.4   Injection Molding and Parameters

Many process parameters, such as the melting temperature, mold temperature, injection pressure, injection velocity, injection time, packing pressure, packing time, cooling temperature, and cooling time, are found to possibly influence the quality of injection-molded plastic products [42–44].

The molding parameters have a great influence on part quality after the thermo-mechanical properties of mold material [45]. An unsuitable process parameter setting can cause many product defects (e.g., a long lead time, a large amount of scrap material, etc.) and unstable product quality during the injection molding process.

Processing parameters have to be set according to the mold cavity design and its size, materials properties, and the quality of molded product without defects. Parameters are a focus since an optimal processing parameter design could help to solve most quality control problems.

4.2.4.1 Temperature

For materials with higher melting temperature, the design and process changes need to be considered. Melt temperature settings reflect the material degradation and are important to achieve the best molded part quality.

The barrel of the injection molding machine supports the reciprocating plasticizing screw. It is heated by the electric heater bands. Increasing heat profile across the barrel may not be the most appropriate profile for all parts. It depends on the size of product and mold; the heat profile may be a reverse, hump, or flat profile. Some adjustments may be necessary to obtain the optimum process conditions and part quality.

Mold temperature is used to facilitate processing and provide good and consistent quality parts at minimum cycle time. To assure good quality molded parts, the actual mold temperature and uniform temperature is extremely important. This can also be done by increasing the number of cavities or using smaller capacity machine relative to shot size, decreasing overall cycle time.

Mold temperature is to reduce stresses in the part and increase in surface gloss. Mold temperature has to be 40–50°C or more. Higher mold temperatures may be desirable for improving part appearance such as higher gloss in injection molding. A higher mold temperature produces a higher gloss and more crystallization. It gives better adhesion in weld lines and increased flow length, but long cooling times. Insulation plates between the mold and clamping plates is at higher temperature than mold temperature. The average cavity temperature is higher than the temperature of the coolant. A temperature variation in one section of the mold surface also can create a non-uniform flow front. The average cavity surface temperature is higher than the temperature of the coolant.

The mold temperature is to balance between part quality and productivity. Higher mold temperature is required on parts that have a long flow path and/or thin wall sections. Higher mold temperature is also required to give good surface appearance with a minimum of warpage.

The average cavity surface temperature is higher than the temperature of the coolant. Thus, set the coolant temperature to be 10–20°C lower than the required mold temperature.

The combined effect of high temperature and high shear rate reduces the melt viscosity, and therefore offsets the pressure requirement. However, high speed generates viscous heating that raises the material temperature.

Longer fill times result in a thicker frozen (solidification) layer, which translates into a more restrictive flow channel, requiring higher injection pressure. The combined effect of high temperature and high shear rate reduces the melt viscosity, and therefore offsets the pressure requirement. However, high speed generates viscous heating that raises the material temperature.

Shorter fill time results in higher volumetric flow rate, and higher pressure requirements.

Longer fill times result in a thicker frozen (solidification) layer, which translates into a more restrictive flow channel, requiring higher injection pressure.

4.2.4.2 Pressure

Three different pressures can be distinguished in the injection unit: injection, hold, back [46]. Pressure is required by both injection unit and clamping system. This force in tonnage should be high enough to prevent the mold from material injection at maximum pressure and speed. The applied force is proportionally related to the projected area of a part including the areas of runner and sprue. It aids the holding pressure and allows for the part to fill and pack properly.

Back pressure should be kept in the optimum range except when additional mixing is required or residence times are already excessive. The growing volume of material in front of the screw will start pushing the screw back once pressure within the melt (i.e., back pressure) is built up to a preset level. Back pressure is the maximum pressure applied during the filling of the mold cavity. Back pressure should be kept in the optimum range except when additional mixing is required or residence times are already excessive.

Low back pressure will result in plasticizing problems and an in-homogenous melt. Higher back pressure will improve the material melt homogeneity and also dispersion of color master batches or other added additives. But the plasticizing time becomes long.

To achieve desired injection speed, it is necessary to have enough injection pressure. Higher injection pressures will improve the weld line strength and at the same time may lead to flashing.

Hold pressure is to promote more chain entangling at the weld line to raise the mold temperature. Increasing pack or hold pressure helps to eliminate low pressure conditions at the weld line. A high holding pressure will improve the adhesion in weld lines and adhesion to inserts or other materials in a multi-shot process. To lower the shrinkage, long holding time is required. Excessive high hold pressures can lead to deformations. Hold pressure adjustment eliminates the problem of flashing.

Higher injection pressure increases

- Wall thickness decreases
- More wall cooling and drag forces
- Long flow length

Lower injection pressure

- Higher wall thickness
- Less wall cooling and drag force
- Short flow length
- Optimal cooling time
- Specifications

The high hold pressure up to a limit without introducing cracks reduces the occurrence of flaws and improves the strength. The wall thickness of plastic parts become thinner; the injection molding operation becomes more difficult. However, the industry demand is for techniques to produce plastic parts with thin wall features.

4.2.4.3   Time

Screw speed influences the residence time of the powder inside the barrel and torque produced during processing. Residence time of material inside the barrel is reduced by the increase of screw speed. Decrease in residence time is due to conveying capacity of rapidly moving screws. Residence time increases with low material feed and torque decrease, due to restricted packing of material. It leads to porous products or the material degrades inside the barrel.

Shorter fill time results in higher volumetric flow rate and requires pressure to fill the mold. Longer fill times require high injection pressure. It results in a thicker frozen solidified layer. Normally, hold time should be as low as possible. Thick parts require long holding times. Short hold time is required for thin parts. Figure 4 illustrates the typical cycle for an injection molding process.

Fill rate and pressure are critical parameters for molding quality parts. Thicker sections of the part fill preferentially due to lower melt pressures required to fill. Plastic flow will accelerate in thicker sections and hesitate in a thin section. The time is a function of the material, volume of part, and mold.

The ejection time is dependent on the hardness of the grade, part, and ejector design. Blade ejectors are used to enlarge the

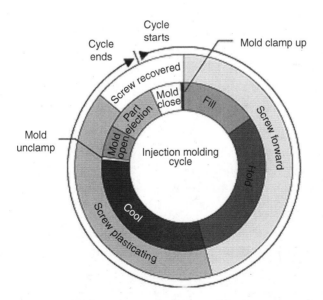

Figure 4. Typical cycle for an injection molding process [Reproduced with permission from Lim, L. T., Auras, R., Rubino, M., *Progress in Polymer Science*, Volume 33, Issue 8, 2008, Pages 820–852 © 2008 Elsevier B.V. All rights reserved]. Ref. [47].

ejecting surface and can reduce the time before ejection. Air assisted part ejection is often used to eject parts with under-cuts. Poppet valves are used in the mold cavity to control the air ejection.

Required ejection forces have a direct impact on mold dura-bility and part quality, and the friction between the mold core and the molded part directly affects the ejection forces [48].

The cycle comprises three stages

- Feeding (screw rotating and moving backwards)
- Stop (no screw movement)
- Injection (screw moving forward without rotation)

The entire process is cyclic, with cycle times ranging between 10 and 100 seconds, depending on the product size and cooling time [49].

4.2.4.4   Cooling

During cooling, location near the cooling channel experi-ences more cooling. The temperature difference between the hot material and mold is to experience differential shrinkage causing thermal stresses. Significant thermal stress can cause warpage problems [50].

On cooling, the relaxation time increases and becomes com-parable to the process time. Part shrinkage decreases with the rate of cooling. In injection molding, mold temperature is an important parameter and related to cycle time of the processing.

To increase productivity in injection molding, minimization of the cooling time is required. Minimization of the cooling leads to parts with higher ejection temperature and poorer mechanical properties. The cooling time depends on the wall thickness of the part, material, melt temperature, and part complexity. It directly affects the productivity of the process and the quality of the product. Cooling time signifies the rate of cooling and economics of the processing cycle.

In injection molding, the mold cooling time usually takes up about 70% to 80% of the time of the entire cycle. Figure 5 shows schematic representation of cycle time in the injection molding process. The relationship between mold cooling time and a molding cycle is directly proportional [51].

Cooling time is the time required to cool the part from its injection temperature to the temperature at which the part is removed from the mold. Cooling time is the major time requirement in the injection molding cycle. Cooling beyond the actual time required will decrease the productivity.

In the injection molding process, during the dwell time the cooling and solidification of the melt starts in the mold. To increase productivity, minimization of the cooling time is required. However, minimization of the cooling time leads to parts with higher ejection temperature and poor mechanical properties.

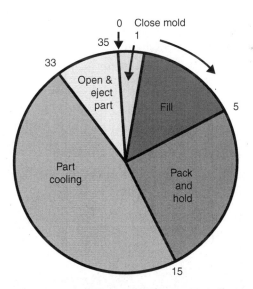

Figure 5. Cycle time in injection molding [Reproduced with permission from D.E. Dimla et al./*Journal of Materials Processing Technology* 164–165 (2005) 1294–1300. © 2005 Elsevier B.V. All rights reserved]. Ref. [52].

The cooling time of the polymer melt in injection molding is a major part of the total time of the processing cycle [53]. Higher mold temperatures will increase cooling time and hence the productivity will be reduced.

4.2.4.5   Velocity

Polymer melt viscosity is very dependent on shear rate [54–57]. Increase the injection velocity, decrease the fill time, and increase the shear rate, which can lower the viscosity of the polymer during fill and thereby allow for better chain entanglement and better packing.

Polymer viscosity is influenced by four main factors: shear rate, temperature, molecular weight, and pressure [58, 59].

Variable orientation as a result of variable melt front velocity with the part leads to differential shrinkage and the part gets warped. It is desirable to maintain a constant velocity at the melt front. Non-uniform velocity uses variable injection speed. Figure 6 illustrates the flow of polymer melt in melt front.

Injection velocity has been recognized as a key variable in thermoplastic injection molding. With the highest possible injection velocity, it has less flow resistance, longer flow length, and improved strength in weld lines. However, it may need to

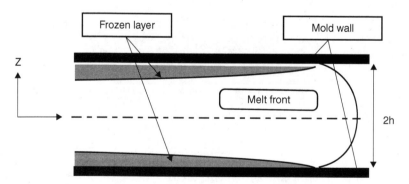

Figure 6. Cross sectional view of melt front. (Reproduced with permission from Seow, L.W., Lam, Y.C.,: *Journal of Materials Processing Technology* 72 (1997) 333–341. © 1997 Elsevier Science S.A. All rights reserved.) Ref. [60].

create vents. Injection pressure is set to the machine maximum to completely exploit the injection velocity of the machine so that the pressure setting valve does not limit the velocity.

4.2.4.6 Part Design

Part design matters because non-uniform wall thickness can vary the shear and flow rate of the melt front, resulting in a split flow path. Thicker sections of the part fill preferentially due to lower melt pressures required to fill. Plastic flow will accelerate in thicker sections and hesitate in a thin section. Sometimes it may be necessary to increase injection rates and round the edge or taper the junction between areas of different thickness. The best answer would be to redesign the part with uniform thickness. A better understanding of the friction conditions during the molding process can lead to improved injection mold and part designs [61].

In the part, ribs increase the rigidity without increasing wall thickness and help to facilitate flow during molding. Ribs are used for load bearing, spacing, supporting, stiffening, guiding, or for restraining.

4.2.5 Injection Molding – Processing

Injection molding is a highly creative manufacturing process [15]. In one automated process, injection molding can be used to make complex geometries in one production step. Machine setup, mold and part design, and selection of material can lead to high quality product. The majority of plastic products are manufactured by the injection molding technique.

Injection molding of thermoplastics involves injecting molten polymer into a mold at high pressure. During the injection phase, the melt material is driven into the mold impression. Upon cooling, the polymer surface tends to replicate the superficial texture of the mold surface.

Five stages of injection molding that repeatedly pave the ways are as illustrated in Figure 7.

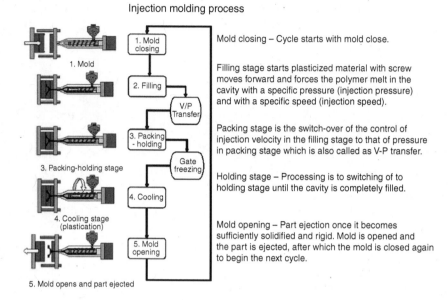

Injection molding process

Mold closing – Cycle starts with mold close.

Filling stage starts plasticized material with screw moves forward and forces the polymer melt in the cavity with a specific pressure (injection pressure) and with a specific speed (injection speed).

Packing stage is the switch-over of the control of injection velocity in the filling stage to that of pressure in packing stage which is also called as V-P transfer.

Holding stage – Processing is to switching of to holding stage until the cavity is completely filled.

Mold opening – Part ejection once it becomes sufficiently solidified and rigid. Mold is opened and the part is ejected, after which the mold is closed again to begin the next cycle.

Figure 7. Illustrates schematic flow diagram of the injection molding process. [Reproduced with permission from Wong, H.Y., *et al. Sensors and Actuators A* 141 (2008) 712–722. © 2007 Elsevier B.V. All rights reserved.) Ref. [62].

Injection molding is to be considered a cyclic process and not a continuous process like extrusion.

Shutdown of an injection molding machine is undesirable, but occurs from time to time in any production setting. In such shutdown, molten polymer is trapped in the barrel of the molding machine, and thermal and oxidative degradation may accelerate and lead to severely degraded polymer and poor part properties. In short stoppages, it may not be an issue for some polymers such as polyethylene; the shutdowns are a concern for polymers containing residual unsaturation, like ABS [63].

4.2.6   Process Variables

Injection molding has machine, process, and quality variables [64]. Figure 8 illustrates the factors influencing product quality and variables involved in it.

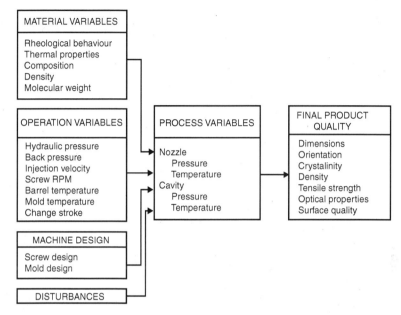

Figure 8. Factors influencing the product quality (Reproduced with permission from Wong, H.Y. *et al. Sensors and Actuators A* 141 (2008) 712–722. © 2007 Elsevier B.V. All rights reserved.). Ref. [62].

Process variables are dependent variables and the collective effect of the machine setting such as cavity and nozzle pressure, melt viscosity, and mold separation [65]. They are

1. Melt temperature (in the nozzle, runner, or mold cavity)
2. Melt front adjustment
3. Maximum shear stress
4. Rate of heat dissipation and cooling.

In injection molding control, along with cavity pressure, nozzle pressure, melt temperature, melt viscosity, and mold separation are widely employed. Process variables are dependent upon the collective effect of the machine setting. Process variables are melt temperature, melt pressure, melt front advancement, shear stress, and the rate of heat dissipation and cooling.

4.2.6.1   Cushion

A cushion of material should remain in the barrel, in front of the screw, after the injection is completed and the mold is filled. The cushion helps to ensure enough material is left for packing the mold. Prevent metal to metal contact during the injection cycle to minimize wear on the nozzle, screw, and barrel. The cushion should contain adequate material for post filling the part, and to prevent the screw from hitting the nozzle. An insufficient cushion may cause sink marks on the part.

The ram position where the filling stage is switched to the packing or holding compensates for volume due to cooling stage. The cushion should contain adequate material for post filling the part, and to prevent the screw from hitting the nozzle.

4.2.6.2   Shot Size

The first stage should fill most of the part but not begin packing. Short size is to confirm the cutoff or switch-over position from first to second state (fill to pack). Short size helps to evaluate the viscosity of materials for applying appropriate second-stage pressure. It also helps to troubleshoot flash, shorts, sinks, bubbles, and splay and to calculate pressure loss over the flow path in filling. Short shot helps to check non-return valve function and the machine function for minimum hold pressure [66, 67]. Short shots are a critical aspect of production process to get identical parts on different machines.

4.2.7   Advantages

Injection molding has many advantages, such as short product cycle, excellent surface of the product, and easily molded complicated shapes. Presently, it is well used in the plastics industry. It also offers many unique advantages for the mass production of small and complex parts. However, the injection molding operation becomes more difficult as the wall thickness of plastic parts becomes thinner [68]. Injection molding helps to produce the micrometer tolerance precision parts of the order of gram mass to tens of kilogram moldings. The injection molding process offers a scope of automatic operations.

4.2.8 Shortcomings

- Skill is required in setting up and controlling the injection molding process
- Machines and molds are more expensive
- Is not tolerant to stop and start conditions of operations
- Injection grade materials are required to control closer tolerances than other grades
- Often requires care to minimize the weld lines
- Irrecoverable loss of materials in sprues and runners
- Labor intensive process

Today injection molding control is accompanied by several key technologies such as novel sensors and information technology [69].

4.3 Extrusion

Extrusion is the most used and popular processing technique for pipes, film, sheet, and profiles. It is a process of converting a raw material into a product of uniform shape and density by forcing it through a die under controlled conditions [70]. It is a continuous process of manufacturing products used mainly to produce different cross-sectional sections that are used mainly in pipe, profiles used in constructions like windows, doors and frames, and pre-fabricated vehicle and aircraft parts and structures. Extrusion plays an important role in plastics processing.

In extrusion, the material pumped through the open channel of a screw extrusion is shared, mixed, and compressed through a die. It is a very essential processing technique to manufacture continuous longitudinal oriented products. Material and flow characteristics are essential to bring out the results of the extrusion process [71–73]. The extrusion process internally is a flow of melt and their interactions. Extrusion is a process of converting a raw material into a product of uniform shape and density by forcing it through a die under controlled conditions [74].

Processing by extrusion, as well as by any other further processing step, plays an important role in polymer lifetime. During extrusion, material is subjected to a kind of energetic shock in which plastic is exposed to elevated temperatures and mechanical stress [75]. In plastics extrusion processes, polymer resin is mixed with several additives at specific ratios to obtain the desired physical and mechanical properties during and after processing. Extrusion process plays an important role and extrusion is the most extensively used and most promising method for polymer processing. Numerous efforts have been made to improve extrusion processing aimed at higher productivity and better product quality [76–80]. Extrusion is suitable for fabricating a variety of shapes with constant cross section, not only for flat shapes, but also for structural shapes such as pipes and tubes [81].

4.3.1   Extrusion – Basic Requirements

Extrusion is the most commonly employed process for fabricating plastic parts. Product quality and production rate are the two most important aspects of the fabrication process. These are primarily a function of the polymer and the extrusion parameters including screw design [82–85].

Plastic extrusion is when the material on its way through the open channel of a screw extrusion is sheared, mixed, and compressed through a die. To manufacture longitudinally oriented products, extrusion is a very essential processing technique due to its continuous production process. Material and the flow characteristics are primarily essential to bring out the results of the extrusion process [71].

There are three types of mixing operations in extruders which can be distinguished as

- Longitudinal mixing or coarse mixing
- Dispersive mixing or continuation of the disperse phase
- Radial mixing or distributive mixing.

In general, the basic requirements of the extrusion process are as follows:

- Constant material quality (bulk density, flowability, rheology, and temperature)
- Low plate-out in die and calibrator
- Low wear of processing unit, die and calibrator
- Melt temperature and pressure measurement
- Constant flow of cooling waters.

In addition to process stability and molecular weight control, output rate is another important consideration. The specific output rate [86] of polypropylene is considerably lower than that of polyethylene of similar shear viscosity. This difference in specific output is attributed to the relatively inefficient melting of PP. Because of this difference in specific output, longer extruders are usually employed for PP commercially.

The output of an extruder almost always exhibits a complex and periodic fluctuation that is the result of the superimposed effect of external disturbances entering the process from various sources and internal flow instabilities. Die flow instability, melting instability, periodic nature of feeder screws, heating elements, non-uniform feed material quality, and the extrusion screw itself are all considered possible causes of process variations.

The response of output flow rate and die pressure is similar to that of a self-leveling system; these parameters return to their original levels after an initial transient period. However, the screw speed can be used to control output flow rates provided that the time period of the input variations is shorter than the transient period mentioned above [87]. Further, for a given screw design, the response is more sluggish requiring better damping at higher percent fills. Deeper screws exhibit better damping ability. The relationship between fill lengths and operating conditions at steady state is nonlinear and depends on the geometry under study [87–88].

Processing ability is influenced not only by the material itself but by a number of other important factors:

- Machine construction (long/short screw extruder, screw geometry, blow head/cooling ring design, blowing direction, winder design)
- Machine quality (functional reliability, measuring and control technology, wear)
- Setting and processing conditions.

4.3.2   Extruder

Extruders are divided into several functional zones, such as solid conveying, melting, melt conveying, and mixing. An extruder will have sizing equipment and occupy the space. An extruder will be using take-off equipment with a cutter synchronized in a manner to have continuous production carried out with product stacking.

The extruder is a pump, and melting is an integral step in plasticating extrusion. To optimize the performance of any kind of pump, it is necessary to have an adequate flow of material through an entry into the equipment. Polymers are melted in the closed barrel system by mechanical energy from the screw rotation and heat transfer from the barrel temperature in screw extruders.

Extrusion process plays an important role in polymer lifetime. Extruder is common equipment in the manufacture of pipe, profile, film, and sheet. The screw design varies with respect to plastic material, melt flow, and the final products. The extruder is divided into several functional zones, such as solid conveying, melting, melt conveying and mixing. In plastic melt, there are three decisive factors. They are screw configuration, screw speed, and feed rate.

Single and twin screw extruders are involved in melting, conveying, compressing, and pumping. These steps can have an influence on the end product quality. Heater bands are strategically located around the cylinder. The plasticating screw pumps uniform melt mix and melt flow through the die.

4.3.2.1   Single Screw Extruder

In a single screw extruder, the plastic material is sheared, mixed, and compressed through a die. The operational differences between the single and twin screw are obvious and each of them will have its own design of modular screw parts that will enhance and improve the mixing effects of the screws.

In extrusion, extrudate swell will depend on the shear rate, shear history, melt temperature, and length to diameter ratio (L/D ratio). In MWD variation it may be advantageous to use polymer whose die swell is particularly responsive or insensitive to die temperature variations.

4.3.2.2   Twin Screw Extruder

The twin screw extruders give better melt temperature control, little dependence on the material friction coefficients, and relatively quick melting. Twin screw extruders will give higher throughput than single screw extruders.

In a twin screw extruder, the screws closely intermesh. In twin screw extrusion, the plastic material taken up from the feeder is conveyed, internally mixed, forced through the gaps and clearance or eventually transferred to the neighboring screw and finally delivers it contents at the die. The material is, in general, enclosed by the neighboring intermeshing screw and barrel inner wall.

The obvious outcome of the mixing efficiency of a screw has been considered to be of major importance. Flow mixing blocks and specific conveying elements help to improve the processing and the output of the machine with better product quality [89].

The twin screw extruder is a versatile and flexible device in plastics processing. It is important to note that in the twin screw extrusion is the distribution of torque input of rotors to every functional zone appropriately.

Twin screw extruders have little sensitivity to die resistance since the throughput is generally defined by an external feeding device (volumetric feeder). The screws and barrel geometries can be tailored. Individual material specifications, characteristics, and output necessity decide the screws and barrel geometries. For certain geometrical configurations such as staggering kneading blocks, or screw with multiple flights, the degree of filling varies along the screw length.

Twin-screw extruders have been of great interest in the processing of complex materials across a range of industries [90]. Among the different classes of twin screw extruders, the intermeshed counter-rotating twin screw extruders are known for better pumping efficiencies combined with low shearing action and narrower residence time distributions [91]. This is a consequence of the axially and radially closed C-cavities formed in an intermeshed counter-rotating twin screw extruder, which progressively move in the axial direction with the passage of time.

Twin screw extruders

- Have additional mixing capabilities.
- More complex
- More expensive
- Flow is 3D and unsteady due to the screw rotation.
- Flow is intrinsically non-isothermal.

In a twin screw extruder, the screw rotates and speeds, the production increases gradually. The effects can be even stronger and they can be rather important to understand, quantify, and control. It is important to know and control rates of ongoing chemical reactions or work with heat sensitive materials [92]. Twin screw extrusion has become an important manufacturing process.

The intermeshed counter-rotating twin screw extruder, with closed channels, acts like a positive displacement pump delivering pressure independent flow [93].

The melt takes different routes through a twin screw extruder. It may require different lengths of time to pass through the extruder. The melt leaving the extruder is called residence time distribution (RTD) [94], which has been used extensively in the literature to characterize the polymer extrusion processes [95–99].

4.3.2.3 Feeder

The size of the feed port may have essentially nothing to do with feeding capacity. The light bulk density powder feeds have to be vented to eliminate the air accompanying the powder. The air will have to flow upstream back to the feed port or to be extruded downstream in the melt.

4.3.2.4 *Screw*

Increasing the lead of the screws does not increase the feeding capacity for light powdery feeds. Plastic material inside a rotating screw is compacted into a solid bed and it melts rubbing on the hot barrel surface. The material forms a thin film between the solid bed, and the metal surface is sheared highly and generates a large amount of heat. During polymer melt extrusion, at high flow rates, polymer melt flow exhibits large oscillations, commonly referred to as spurt [100].

The light bulk density powder feeds have to be vented to eliminate the air accompanying the powder. The air will have to flow upstream back to the feed port or to be extruded downstream in the melt.

The physical and chemical properties drastically change from location to location within the screw length. Different screw designs possess different relationships and are related to the fully and partially filled screw, respectively.

The screw rotation enhances the melting of polymer due to frictional heat generation inside the barrel. The screw turns, melt film adheres to the barrel to be stretched, or sheared. The rotation of screw also continuously scrapes the melt film of the barrel wall, collecting it on the front side of the flight. Even the rotation causes the un-melted resin to be pushed continuously toward the barrel wall. The renewal process allows the polymer to be melted faster without relying solely on conduction heat from the melt film on the barrel wall.

At high screw speeds, the heat generated inside the melt film and barrel usually removes the excess heat from the melt. Plastics with high viscosity generate more heat in extrusion and melt faster. Material with lower heat content and lower melting temperature will melt faster.

There are three kinds of mixing: dispersive, distributive, and extensional. We will deal here only with dispersive and distributive mixing, as extensional mixing occurs predominantly in twin-screw extruders. Dispersive mixing is like putting two materials to be mixed between two plates and rotating one of the plates. The shear stress developed in the polymer between the plates would be proportional to the distance between the plates and the speed at which the plate was rotated.

Operating variables, such as screw speed, barrel temperature set point, are used as forcing functions to disturb the extrusion system. Because of these disturbances, the extrusion process variables, such as motor torque, die pressure, and product temperature, are shifted from their initial steady states to new equilibrium values.

Initial increases of the torque and die pressure as the screw speed is increased indicate the effort of motor drive to overcome the inertia of screw shaft and the momentary increase in the output. Increases in screw speed at constant feed rate would result in decreases in the degree of fill in the barrel since the latter is dependent on the ratio of feed rate and screw speed.

Changes in barrel temperature set point itself involve a dynamically changing input. In other words, the barrel heating or cooling involves dynamics, which are then reflected in the process outputs. Dynamic relationships exist between the operating (input) and process (output) variables of the extrusion process.

4.3.2.5   Die

The swell of the extrudate as it leaves the extrusion die is an important phenomenon in polymer melt extrusion [101].

Die pressure [102, 103] is

- Affected by flow rate of plastic melt
- Cross-section of the die exit

- Die temperature
- Material viscosity

Entrance angle should be optimum to minimize the pressure drop [104]. However, there is no considerable effect on the mechanical strength of the product with the die entrance angle. The land length will have some effects [105].

The following points are important for the design of the die

- Inlet pressure drop at point of restrictions.
- Entrance vortices
- Extrudate swelling after it leaves the die.

4.3.3 Polymer Melt

The twin screw extruder is a versatile and flexible device in plastics processing. It is divided into several functional zones, such as solid conveying, melting, melt conveying, and mixing. It is important to note that in the twin screw extrusion is the distribution of torque input of rotors to every functional zone appropriately.

Plastic melt viscosity is very dependent on shear rate [106–108]. The plastic melt is highly viscous during processing and depends on molecular weight, molecular weight distribution, chain branching, shear rate, and shear stress [109].

Process variations can be caused by die flow and melting instability, feeder screw, heating elements, non-uniform feed material quality, and the extruder screw itself. The output of an extruder always exhibits a complex and periodic fluctuation that result in the superimposed effect of external disturbances entering the process from various sources and internal flow instabilities. Under certain conditions, the amplitudes of these fluctuations are large enough to cause undesirable effects such as surging or spurt flow, bamboo fracture, sharkskin, and other forms of product non-uniformity.

It can be caused by die flow temperature relative to the screw in different stages of reciprocating screw [110]. The rate of solid and melt conveying in reciprocating screw is reduced by the axial movement of the screw velocity [111]. The melting rate reduction in axial screw movement results in a larger melting zone. The melting rates change from time to time as well as the melting lengths.

The polymer melt is highly viscous and depends on molecular weight and molecular weight distribution. It also depends on chain branching, shear rate, and shear stress. Die swell increases with shear rate. Properties increase with molecular weight and molecular weight distribution. However, increase in molecular weight gives processing problems, due to its higher viscosity. Decrease in density increases the stress cracking resistance [112].

4.3.4   Extrudate Swell

Extrusion of a hot melt through a cooler die zone substantially increases the extrudate swell of some thermoplastics. Extrudate swell, or die swell as it is more commonly known, is the property of expansion of viscoelastic fluids on extrusion from a constriction [113–116].

Extrusion of a hot melt through a cooler die zone substantially increases the extrudate swell of some thermoplastics. Extrudate swell, or die swell as it is more commonly known, is the property of expansion of viscoelastic fluids on extrusion from a constriction [117–120]. Extrudate swell is dependent on shear rate [121] or more universally on shear stress [122], polymer molecular weight distribution [123], the length to diameter (L/D) ratio of the die [124], and the difference in diameters of the extrusion reservoir and orifice [125].

The swelling of polymer melts must be taken into account in the design of extrusion and blow molding process. Extrudate swell significantly affects the wall thickness of blow molded products. Higher shear rate increases the extrudate swell [126].

Extrudate swell may be significant in blow molding because it could be possible to control such parameters as product wall thickness by the die temperature alone.

In extrusion, die swell and compliance may decrease or increase depending on whether MW is predominating or whether effect due to broadening of MWD is predominating [127].

There is general agreement that extrudate swell results from relaxation of elastic stresses in the fluid [128]. The swelling of polymer melts must be taken into account in the design of extrusion and blow molding processes.

In extrusion, extrudate swell will depend on the shear history and rate, melt temperature, and length to diameter ratio of the screw. With molecular weight distribution variation, it may be advantageous to use a polymer whose die swell is particularly responsive or unresponsive to die temperature variations.

4.3.5   Extrusion and Process Parameters

In extrusion, the distributing factors are

- Feed to the extruder, e.g., powder or solid material
- Melt index
- Viscosity
- Temperature and stress field.

The temperature history during processing is a critical parameter in dictating final part strength and depends upon the rate at which the extrudate cools upon leaving the extrusion head. The thermal history imposed on the material has important repercussions on the final product.

Most of the melting occurs in the compression or barrier section of a single screw and accounts for approximately 85 to 90% of the drive power requirement. It is important to note that the same stretching or shearing mechanism occurs in the metering section. In the metering section, the polymer melt is equal to the channel depth because most, if not all, the polymer is melted at that point [129].

Presence of moisture in the material in the extruder during extrusion forms steam along with the melt. Back flowing steam will then condense on the entering cold powder feed. It leads to the complete reflux of the moisture. This can lead to bad surging in the extruder. Melt flow properties are useful in selecting an appropriate extruder screw and die, in setting appropriate processing conditions in troubleshooting extrusion problems, and in allowing prediction of extrusion behavior.

Initial increases of the torque and die pressure as the screw speed is increased indicates the effort of motor drive to overcome the inertia of screw shaft and the momentary increase in the output. Increases in screw speed at constant feed rate would result in decreases in the degree of fill in the barrel since the latter is dependent on the ratio of feed rate and screw speed.

Changes in barrel temperature set point itself involve a dynamically changing input. In other words, the barrel heating or cooling involves dynamics, which are then reflected in the process outputs. Too short land length will create flow change with any change in run conditions. Land length helps to offset any swell of the thermoplastic as it exits the die lip.

In extrusion, shrinkage is the combined effect of structural stress accumulation due to stretching and thermal stress accumulation due to constrained cooling. Polymer melts are poor heat conductors and thus most heat conductance will occur in the direction normal to the plane of the product. The stretching stops when the polymer product such as pipe, sheet, film, etc. fully contacts mold. In extrusion, mold may be a calibrator for pipe extrusion, calendar for sheet extrusion, etc. The strain involved in the stage is negligible compared to that in the stretching stage. Die flow instability, melting instability, periodic nature of feeder screws, heating elements, non-uniform feed material quality, and the extrusion screw itself are all considered possible causes of process variations.

A decrease in extrusion pressure is of benefit to dies and extruder. With high performance screws, it is possible to increase the output of an extruder. Process variations are caused by die flow instability, melting instability, periodic nature of feeder screws, heating elements, non-uniform feed material quality, and the extrusion screw.

During shut down, the screw is reduced to zero in very short time. Hence the distributed residual solid plastic material may undergo changes. Being the material is partially filled or the depth of the screw is large, the solid plastic may be melted when the barrel is being quenched. Shut down process is time consuming.

As the volumetric flow rate decreases, the pressure gradient increases and hence the back flow, i.e., negative velocity zone becomes more pronounced. During extrusion the flow in the down channel direction is a combination of drag and pressure flows [130].

4.3.6   Extrusion – Processing

Extrusion is the most commonly employed process for continuous production of plastic parts. Product quality and production rate are the two most important aspects of any fabrication process. These are primarily a function of the polymer and the extrusion parameters including screw design [131–134].

Extrusion is a continual process to make products such as film and sheet, wire covering, pipe, and profiles. Essential components include the hopper, barrel, cylinder, plasticating screw, thrust bearing, breaker plate and screen pack, backpressure regulating valve, die adapter, die, sizing, and haul-off equipment. Heater bands heated by built-in thermocouples are strategically located around the cylinder. Figure 9 illustrates the flow situation in extrusion.

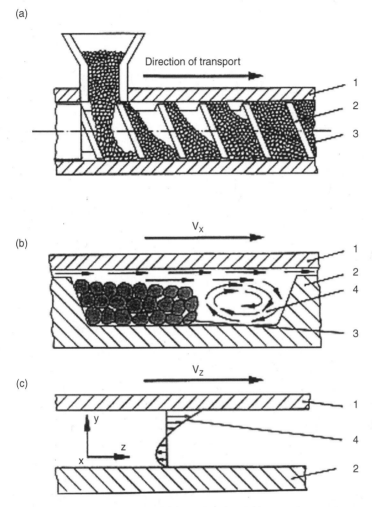

Figure 9. Flow situations in a barrel/screw system. (a) conveying zone
(b) metering zone (c) transition zone. 1. Barrel 2. Screw 3. Solid plastic 4. Melt
(Reproduced with permission from Heinze, M., *Surface and Coatings Technology*
105 (1998) 38–44. © 1998 Elsevier Science S.A. All rights reserved.). Ref. [135].

Variations in molecular weight distribution would be
expected to affect the temperature dependence of melt elasticity. For some applications, it may be advantageous to use a
polymer whose die swell is particularly responsive or unresponsive to die temperature variations.

The screw rotation enhances the melting of polymer due to frictional heat generation inside the barrel. The screw turns, melt film adheres to the barrel to be stretched, or sheared. The screw rotation also continuously scrapes the melt film of the barrel wall, collecting it on the front side of the flight. Even the rotation causes the un-melted resin to be pushed continuously toward the barrel wall. The renewal process allows the polymer to be melted faster without relying solely on conduction heat from the melt film on the barrel wall.

The temperature history during processing is a critical parameter in dictating final part strength and depends upon the rate at which the extrudate cools upon leaving the extrusion head. The thermal history imposed on the material has important repercussions on the final product.

Most of the melting occurs in the compression or barrier section of a single screw and accounts for approximately 85 to 90% of the drive power requirement. It is important to note that the same stretching or shearing mechanism occurs in the metering section. In the metering section, the polymer melt is equal to the channel depth because most, if not all, the polymer is melted at that point [136].

The extrusion process is a series of sub-processes including motor drive control system, barrel heating/cooling system, the combination input, and the extruder and die assembly itself.

The extrusion process can be divided into three major sections: (a) Solids Conveying (b) Melting and Compaction and (c) Metering. The factors that enhance the solids conveying efficiency in the feed section of an extruder are: (i) large screw channel area; (ii) low friction on screw surface; (iii) high friction on barrel surface; and (iv) long feed section. The melting efficiency of an extruder is influenced by the temperature profile, the screw type, the melting characteristics of the polymer, and the viscous heat generated. The primary function of the melt-metering section is to convey a homogenous melt to an appropriate die.

Melting takes place in the extruder and the melt becomes viscous dissipation primarily. The melts result from the rotation of the screw in the barrel. The melt film adheres to the barrel during screw rotation. The melt film is to be wiped from the barrel wall continuously while the screw flights collect it on the front side of the flight. The channels become shallower in the compression section of the screw, causing the unmelted resin to be pushed continuously toward the barrel wall. This renewal process allows the polymer to be melted faster without relying solely on conduction of heat from the melts film on the barrel wall.

During extrusion, material is subjected to a kind of energetic shock of which plastic is exposed to elevated temperatures and mechanical stress. It may lead to a dramatic change in the polymeric structure [137]. The extrusion process internally is a flow of melt and its interactions.

Screw design plays an important role and optimum processing conditions lead to achieving excellent product quality. Extrusion requires plastics material with higher molecular weight for better melt strength. The extrusion process internally is a flow of melt and their interactions.

In extrusion, the material is heated and pumped into a permanent mold which is called as die enclosed in a barrel, through an opening, the product cross-sectional area and dimensions are shaped and cooled. A product of uniform shape and density is obtained under controlled conditions [138].

The actual plastic zones indicate the possibility of localization of plastic deformation. It is also important to obtain a model of the flow field. The formation of the deformation zone depends on the die shape also. For a conical die, dead zone does not exist. For a flat die, a dead zone creates its boundary as a part of the boundary of the plastic zone. However, the optimum angle of the die for co-extrusion does not have the same meaning as for mono-material. A poorly designed die does not permit the part to be extruded with the same dimensions from run to run – coupled with a lack of understanding of the extrusion process, is a recipe of scrap generation.

In plastic processing, with the equipment at some stage, either in single or twin screw, the screw is involved in melting, conveying, compressing and mixing of the material. The steps involved can have an influence on the final process quality.

4.3.7 Advantages

- Continuous process with low labor oriented.
- Higher output
- Ability to achieve an acceptable quality of finished product
- Process control within a narrow fluctuation temperature range
- Obtain the desired shape and properties of the products.

4.3.8 Shortcomings

- High capital investment and more space requirement
- Limited product flexibility
- High initial wastage.

4.4 Blow Molding

Blow molding is one of the important methods of manufacturing hollow plastic articles. The process is traditionally used to produce containers with different sizes.

Blow molding is limited by equipment that is feasible for the mold sizes. Blow molding is the world's third largest plastics processing technique. It is used to produce hollow, thin wall objects from thermoplastic materials. In the last 20 years blow molding has seen a rapid growth due to the development of new application areas in the automotive, sports and leisure, electronics, transportation and packaging industries [139]. The complexity of these new molding techniques calls for a much better understanding of the process, machine, and material behavior and its effect on the performance of the final part.

4.4.1   Blow Molding and Process Parameters

The properties of the resin, thermal and mechanical history are the dictating factors on the quality and dimension of extrusion blow molded articles. The process involves plastication, parison formation, clamping, inflation, and cooling stages. Formation of the parison is one among the critical parts of the process. The dimensions of a blow molded article will be directly related to the dimension of the parison. The material's thermo-mechanical history and parison diameter distribution and the resulting weight influences the phenomena that occur during the subsequent inflation and cooling stages.

Operating parameters – machine setup

- Die gap
- Melt temperature
- Inflation pressure
- Cooling time
- Temperature of the coolant liquid.

Sag in blow molding is best described as the extension of the molten parison due to gravitional force. Any variance in the sag properties of a resin will have a direct effect on the shape and physical characteristics of the bottle. However crude the sag measurement, the observed changes with multiple passes is evidence that some high MW material in the copolymer is lost, whereas some chain lengthening or crosslinking occurs in the homopolymer and natural PCR [140].

4.4.2   Extrusion Blow Molding

Parison formation in extrusion blow molding consists of the extrusion of a polymer melt through a complex annular die. The process involves varying the die gap during the extrusion and therefore results in a parison having variable thickness along its length. Preform conditioning in reheat injection blow molding entails the application of profiled radiative heat onto the preform. The preform can also be of a variable thickness along its length.

The process consists of four steps:

- The extrusion of hollow cylinder of molten parison
- Clamping of the mold halves around the parison
- The blowing of air into the part to expand it against the sides of the mold
- The cooling of the part before part ejection.

Expansion takes place against the sides of the mold, and the cooling of the part before part ejection. Figure 10 illustrates the schematic representation of extrusion blow molding.

Polymer powder or pellets, colorants, and other additives are fed to a rotating screw extruder where they are mixed and heated into a homogenous melt. The plastic melt is forced through a die, which forms the plastic into a parison shaped as a cylindrical tube.

For cooling stage in extrusion blow molding process, the cycle time represents 60% or more. The cooling time directly influences the productivity and cost of the process. Cooling and solidification affect the properties. Excess cooling limits production rate, whereas short cooling results in part deformation.

Cooling stage represents a substantial part of the overall cycle. Cooling stage is to decide the ultimate properties of the molded part. Heat is removed both by forced convection (polymer/air interface at the internal wall) and conduction (polymer/metal

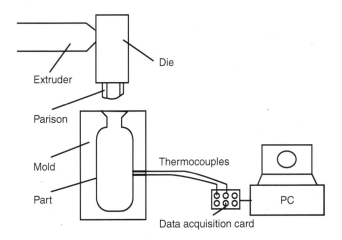

Figure 10. Schematic representation of extrusion blow molding [Reproduced with permission from Huang, H.X., *et al. Polymer Testing* 25 (2006) 839–845 © 2008 Elsevier B.V. All rights reserved]. Ref. [141].

interface at the mold wall). When the mold opens, the part will continue to cool from both surfaces by natural convection.

4.4.3   Injection Stretch Blow Molding

The injection stretch blow molding process is used for manufacturing hollow articles. Injection stretch blow molding offers decisive advantages over extrusion blow molding in regard to mechanical properties, barrier properties, and transparency, especially when using semi-crystalline materials such as polypropylene. For getting a good product quality, the ratio to which the material has to be stretched must be fairly high.

Injection stretch blow molding is used for the mass production of bottles.

It is a two stage blow molding process

1st stage

- Injection molding of pre-form

2nd stage

- Preheating of pre-form uniformly to a pliable state
- Blowing stage of the process
- Pre-form in place of the conventional infrared oven heating.

Infrared heating provides low energy efficiency for heating of plastics and the heating time is high.

The injection molded preform is placed into the machine and heated up from room to stretching temperature by IR heaters. Then the parison is moved into the mold, axially stretched by a plug, and blown. When the cooling time has passed, the article is demolded. When the heating is finished, the so called transient time is passing before the actual stretching starts. The transient time is supposed to level the temperature profile across the preform wall and also allows reading the preform surface temperature simultaneously at different spots across the preform lengths by a device especially developed for that purpose.

In stretch blow molding the stretching temperature has the most significant influence on the article performance besides the stretching ratio. The processing window for the stretching of the used PP homopolymer is within a temperature range of 145–153°C [142].

The subsequent inflation for the parison or the preform, is a complex phenomenon involving the interaction of many process variables, especially the initial thickness distribution and the melt temperature [143–145]. The optimal final part thickness distribution can therefore be enacted by (i) manipulating the parison or preform thickness distribution and/or (ii) manipulating the parison or preform melt temperature. Therefore, the design of blow molded parts calls for the optimization of parison or preform conditions.

The main limitations of this infrared heating are that it provides low energy efficiency for the polymer heating and the

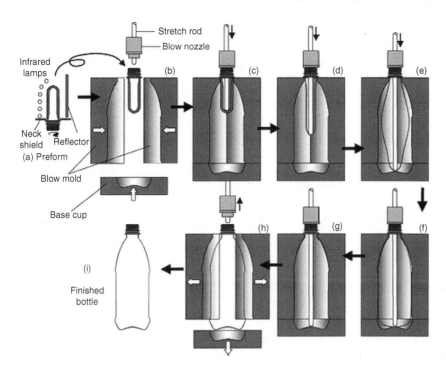

Figure 11. Injection stretch blow molding [Reproduced with permission from Lim, L..T., Auras, R., Rubino, M., *Progress in Polymer Science*, Volume 33, Issue 8, 2008, Pages 820–852 © 2008 Elsevier B.V. All rights reserved]. Ref. [47].

heating time is also high (of the order of 25 s in a typical process). Microwave heating technology is seen as potentially more effective in terms of power consumption and process time. Dielectric characterization of PET perform is important to subject to the microwave power. Microwave heating is different from conventional heating. The dielectric constant and loss factor give information about the heating rate, penetration depth, and the microwave absorption [146].

Stretch blow molding is used for manufacturing hollow articles. Stretch blow molding process offers advantages of products with improved mechanical properties, barrier properties, and transparency, especially when using semi-crystalline material such as polypropylene.

4.4.4   Advantages

- Process for low to high volume containers and hollow parts
- Possible to produce wide range designed containers.

4.4.5   Shortcomings

- Labor oriented processing
- Dimensional tolerances are difficult to achieve
- Product with low surface finish
- Always with possibility of parting lines.

4.5   Thermoforming

When the polymer sheet is heated above glass transition temperature, it behaves more viscously than a hyper-elastic bubble. The stress relaxation, which occurs during the entire forming process, is also heavily related to the temperature effect [147].

Thermoforming process is a combination of a complex series of events, such as the clamping and removal of the polymer sheet, the shuttling of sheet from station to station, the open and closing of the mold platens, the exertion of pressure and vacuum, the air-assist part ejection, and the on/off of the cooling fans. [148].

The thermoforming process continues in the industry due to its low cost and good formability. It is most widely used in packaging industries. Other applications include making large parts such as refrigerator door liners, bathtubs, signs, and automotive interior trim.

The molecular orientation resulting from blow molding and thermoforming can greatly improve the mechanical properties of the formed product. The end result is a well proportioned part (bottle, for example) with appropriate thickness and weight distributions.

Thermoforming has become a major process with a growing range of applications in the automotive, aerospace, agricultural, building, and packaging industries [149–150].

4.5.1   Thermoforming and Parameters

Product consistency from thermoforming depends on controlling variables of material and process. The important variable is the quality of the sheet feed stock. The extruded sheet of the same material could vary in terms of polymeric contamination, thickness, thermal stresses, and amount of regrind, volatiles, color, gloss, and grain. Variation in mold material and mold temperature also affects the product consistency. Obviously during actual thermoforming, many times the effects of such variations results in tearing, wall thinning, shape distortion, fading, pinholes, and grain distortion [151].

The term thermoforming defines a set of techniques which allow the production of thin wall plastic shapes starting from flat sheets or films. In the process, the polymeric sheet is clamped and heated above its glass transition [Tg] or melting temperature [Tm], where the material is softened, has a marked visco-elastic behavior, and can be easily formed. The sheet is then placed on a metal mold and forced to assume the desired shape by the help of vacuum or air pressure. After cooling, the part is refined. Typical applications of this technology are the productions of containers, appliances cells and covers, vehicles internal panels, etc. made of different materials, such as ABS, HDPE, PVC, PP, PS and HIPS.

In general, the suitability of thin-walled thermoformed parts in tight tolerance applications will be limited by the linear shrinkage of the part relative to required tolerances. While out-of-plane warpage is significant, thermoformed parts are highly compliant and are easily straightened into the required shape [152–153].

4.5.2   Processing

Thermoforming is a process by which a plastic sheet, made soft and flowable by the application of heat, is biaxially deformed into a three-dimensional shape by applying either vacuum or pressure, or both [154–155].

Thermoforming is a process with polymeric sheet deformed inside the mold under vacuum. The sheet is heated above glass transition temperature; it behaves more viscously than a hyper-elastic bubble. Other variant processes involve stamping of sheet-polymer parts by means of punches and dies [156].

Thermoforming is a widely used process to manufacture thin-wall polymer parts at relatively low cost. In the process, an exerted pressure differential rapidly deforms the softened polymer sheet to fully contact the mold surface. The formed part is usually held against the mold surface by a pressure differential to facilitate cooling and reduce final shrinkage, after which it becomes capable of sustaining the newly acquired shape. The part is then ejected and trimmed into the final product.

4.5.3   Mold

The mold temperature setting depends on the design of the mold. Tool with bar locks, the setting of the mold temperature appropriately to provide the best forming results. In case there is an undercut on the tool, a mold temperature should allow the formed product to shrink a bit and eject from the mold easier.

In the mold, the following points are essential

- Minimize the wall thickness as possible
- Use large radii
- Stiffening by curvature instead of rib
- Simple mold partitioning
- Sufficient tapering to easilly eject a product

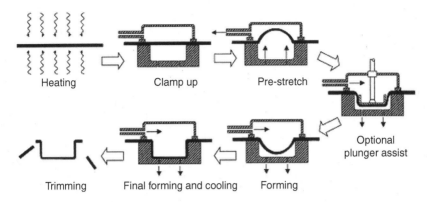

Figure 12. Main steps for thermoforming [Reproduced with permission from Lim, L.T., Auras, R., Rubino, M., *Progress in Polymer Science*, Volume 33, Issue 8, 2008, Pages 820–852 © 2008 Elsevier B.V. All rights reserved]. Ref. [47].

- Avoid scratching of etched surfaces
- Avoid undercuts
- Avoid weak spots in molds and products.

The mold temperature setting depends on the design of the mold. Tool with bar locks, the setting of the mold temperature appropriately to provide the best forming results. In case there is an undercut on the tool, a mold temperature should allow the formed product to shrink a bit and eject from the mold easier.

4.5.4   Advantages

Thermoforming has advantages over its better known competitor processes such as injection molding and compression molding, because it uses simpler molds and a much lower forming pressure. Thermoforming is the process of choice where short production runs cannot justify the expense of the more expensive injection tooling, or where short lead times from design to production are critical. Larger parts like bathtubs or refrigerator door liners are only economically possible by thermoforming.

4.5.5 Shortcomings

- Requires material in sheet or film
- Temperature of the sheet or film varies within the oven
- Wastage of material occurs during shaping operation.

4.6 Rotational Molding

Rotational molding, also known as rotomolding is a shear and pressure free process used to manufacture hollow plastic parts such as toys, containers, tanks, road sign hollow partition, and playground equipment with relatively low investment [156].

Rotational molding is subjected to heating and then cooling. Semi-crystalline thermoplastic materials are mostly used. Material selection is based on toughness and stiffness. The particle size of the powdered material is required typically of 35mesh particle size, in a biaxially rotating mold. It involves the tumbling, heating, and melting of thermoplastic powder, typically of 35mesh particle size, in a biaxially rotating mold, followed by coalescence, fusion, or sintering and cooling. Machine design plays an important part in the manufacturing of quality molded products [157–160].

Polyethylene tends to dominate the market as a rotomolding material due to its high thermal stability and wide range of grades that have been developed specifically for the process. However, other thermoplastics such as PVC, ABS, PC, polyester, polyamides, and even some thermosets can be rotomolded successfully [161–164] to assist with the identification of optimum speed ratios, temperatures, etc. Machine design also has an important role to play in the production of good quality moldings.

4.6.1 Rotational Molding and Parameters

The fundamental and controlling stages in the rotational molding cycle are the powder melting and sintering [165].

Inside the mold, melt viscosity has significant influence in rotational molding. Particle size distribution affects the porosity of the product [166]. At optimum temperature, the density of the sintered product is maximum for a good quality rotational molded product [167].

The mold material and its internal surface finish are two of the most important factors that influence the appearance of a product. Internal air temperature enables part properties to be optimized with processing conditions.

Rotational molding with higher particle sizes or high viscosity materials results in poor surface reproduction and tendency to trap air in the form of bubbles when the part is formed [168]. At very long cycle times, the molded part will have no bubbles. However, impact strength is low because the PE has oxidized.

Impact strength is an important mechanical property, as most service failure occurs when a product is dropped or has something dropped on it. However impact resistance is probably one of the least understood mechanical properties of the polymer. The impact strength of the final product is of great importance in rotational molding. Impact strength includes not only the molecular structure of the polymer, but also the shape and size of the material and processing conditions applied while manufacturing the products [169].

Inside the mold, melt viscosity has significant influence in rotational molding. Particle size distribution affects the porosity of the product [164]. Higher rotation speed can improve the thickness distribution of rotationally molded parts. As the temperature rises in the mold, the powder softens and melts, adhering to the mold wall [169] and forming a homogenous, bubble free melt pool along the entire inside surface of the mold. Biaxial rotation ensures that the powder is evenly distributed in the mold. The amount of powder determines the wall thickness of the rotomolded part.

During the rotational molding process the mold is rotated at relatively slow speeds. Unlike other polymer processing techniques this causes low shearing forces between the molten polymer and the mold wall.

4.6.2 Mold

The quality of any rotationally molded item is governed principally by the quality of the mold used. Typically, molds are made from mild steel, stainless steel, cast aluminum, or electroformed nickel/copper. The choice of mold material depends on the particular application, the number of molds required, and the surface finish necessary. Another important factor is the type of release agent that is used. Release agent is necessary to aid part removal.

Rotational molding is restricted to low shear rates because the mold must rotate slowly, to minimize the inertia forces generated as the relatively heavy mold and contents are rotated.

Some molds produce parts that contain areas with no surface pores, while other areas of the same molding have surface pores. It is due to hot spots or areas of the mold that conduct more heat [170].

4.6.3 Processing

The plastic powder is placed in one half of the mold portion. The mold is then closed and subjected to biaxial rotation in an oven with required processing temperature. The plastic powder inside the mold is melted by heat transferred through the mold wall. After all of the powder has melted, the mold is moved out of the oven, while biaxial rotation continues. Still air, a blowing fan, or a water shower is usually used to cool the mold. Once the product inside the mold is cooled to a state of sufficient rigidity, the mold opens and the product is removed [171–174].

During the tumbling process, the finer powder particles get sieved down closer to the wall and larger particles form layers on top. As the temperature rises in the mold, the powder softens and melts, adhering to the mold wall [175] and forming a homogenous, bubble free melt pool along the entire inside surface of the mold. Biaxial rotation ensures that the powder is evenly distributed in the mold. After the heating cycle is completed, the mold is cooled resulting in solidification of the polymer. The amount of powder determines the wall thickness of the rotomolded part.

Rotational molding process involves four steps [176–179]. They are:

1. In the separable cast or fabricated vented mold, pre-determined and weighed amount of powdered plastic material is charged;
2. The powdered material is heated in an oven with biaxial rotation and external heating without applying pressure or centrifugal force until powdered plastic melts and coats in the internal surface of the mold;
3. After all of the powder has melted, the mold is moved out of the oven, while biaxial rotation continues [180–184]. Rotating mold is cooled externally with forced air or water mist to allow the molding to solidify;
4. Removing the part from the mold's cavity.

Figure 13 illustrates the principle of rotational molding of plastics. Rotational molding is the process of producing hollow parts by adding plastic powder to a shell-like mold and rotating the mold about two axes while heating the mold and powder.

The fundamental and controlling stages in the rotational molding cycle are the powder melting and sintering [185]. It involves the tumbling, heating, and melting of thermoplastic powder, typically of 35mesh particle size, in a biaxially rotating mold, followed by coalescence, fusion or sintering and cooling. [186–189].

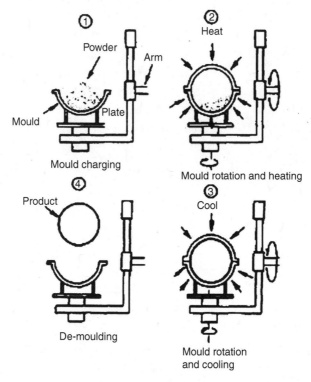

Figure 13. Principles of rotation molding of plastics [Reproduced with permission from Crawford, R.J., *Journal of material processing technology*, 56 (1996) 263–271 © 2008 Elsevier B.V. All rights reserved]. Ref. [183].

4.6.4   Pigmentation

With the exception of carbon black, the inclusion of compounded pigmentation does not affect the rate at which air bubbles diffuse out of the polymer melt. However, the inclusion of carbon black pigmentation was shown to increase the viscosity of the material, and also inhibit the diffusion of air bubbles out of the polymer. The shape, size, and particle size distribution of powder particles used during rotational molding have been shown to influence the size and density of bubbles. In general, increasing the number of finer particles produces more bubbles that are physically smaller. It has also been highlighted that finer particles tend to sieve through the coarse particles during rotational molding, and melt first against the wall of the mold.

4.6.5   Advantages

- With production capacity, rotational molding requires low investment
- Complex shapes can be manufactured without any need of post assembly
- Possible to produce two or more parts
- Easy to change color and material
- Minimum wastage.

4.6.6   Shortcomings

- Not ideal for producing parts with holes through the wall
- Long cycle time
- Difficult to monitor and control the melt
- Products with low mechanical and shock mitigation properties due to the hollow structure [190]
- Wall thickness cannot be controlled precisely.

4.7   Fundamentals

4.7.1   Injection Molding

- Complete injection molding process cycle comprises mold closing phase, injection of melt into cavity, packing pressure phase for compensating shrinkage effect, cooling phase, mold opening phase and part ejection phase
- Plastic part is the combination of proper part design, material selection, tool design, and optimized process

- The injection and extrusion process uses screws to convey, pump, and blend the heterogeneous components. Extrusion produces continuous linear profiles by forcing a melted thermoplastic through a die. The injection molding process produces three-dimensional items
- Cost reduction and quality improvement are the main driving forces in the injection molding industry
- The time is a function of the material, volume of part and mold
- The cooling time depends on the wall thickness of the part, material, melt temperature, part complexity
- Air assisted part ejection is often used to eject parts with undercuts.

4.7.2   Extrusion

- Material and the flow characteristics are primarily essential to bring out the results of the extrusion process
- Screw extrusion is sheared, mixed, and compressed through a die
- Extrusion is a very essential processing technique due to its continuous production process
- Poorly designed die does not permit the part to be extruded
- Extrusion is suitable to fabricate variety of shapes with constant cross section
- In the closed barrel system, the plastic is melted by mechanical energy from rotating screw and heat transfer from the high temperature barrels
- Plastic melt viscosity is very dependent on shear rate.

4.7.3   Blow Molding

- Blow molding is a continuous process capable of producing high production rates

- The extrusion blow molding process involves three main stages, i.e., parison formation, parison inflation, and part solidification
- Product development is defined as the time required bringing the design of the part and process to an optimal condition
- Factors affecting the volume and mass of the produced bottles are cooling time, screw speed, melting temperature, blow pressure, blowing time, and mold temperature
- Factors related to mold behavior are mold opening time, closing time of the mold, closing speed, and mold position
- Factors to determine the part produced are blow pressure, screw speed, cooling time, and melt temperature
- The ratio of the material to be stretched must be fairly high to have good product quality.

4.7.4   Thermoforming

- The quality of the thermoformed parts depends on many processing variables such as temperature of heating pipes, vacuum pressure, and plugs material, plug moving speed and plug displacement
- Shrinkage is the combined effect of structural stress accumulation due to stretching and thermal stress accumulation due to constrained cooling
- Tight tolerance application will be limited by the linear shrinkage of the part relative to required tolerances
- Good melt strength is required as the material is formed over an open mold
- Prior to thermoforming, it is important to establish a starting point for the heating platen temperature to minimize the wastage of sheet materials

- During thermoforming, the sheet thins which makes it necessary to optimize the process before molding a part [191–194]
- Polymeric materials are poor heat conductors, and thus most heat conductance will occur in the direction normal to the plane of the sheet
- Thermoforming is a highly complex, multivariable, and non-linear process [195].

4.7.5 Rotational Molding

- Rotational molding is used to produce relatively large sized parts, hollow, complex shaped and single piece plastic articles
- Heating, shaping, and cooling of the plastic all takes place inside the mold without application of pressure
- Rotational molding is a zero shear process

Heat transfer in rotational molding assumes that the heat transfer at the mold powder interface is because of convection, whereas the powder particles are heated up by conduction.

References

1. Tadmor, Z., The evolution of polymer processing into macromolecular engineering and science, *Plastics, Rubbers and Composites*, 2004 vol.33 No. 1, 3–4.
2. Yang, H., Zhan, M., Liu, Y.L., Xian, F.J., Sun, Z.C., Lin, Y., Zhang, X.G., Some advanced plastic processing technologies and their numerical simulation, Journal of *Materials Processing Technology*, 151 (2004) 63–69.
3. Macosko, C.W., *Rheology: principles, measurements and applications*, New York: VCH Publishers, 1994.
4. Isayev, A.I., Guo, X., Guo, L., Demiray, M., Microstructure of injection moldings of isotactic polypropylenes with various molecular weights: simulation and experiment. *Annual Tech Conf., Soc. Plast. Eng.*, 1997;2(55):1517–21.
5. Angstadt, D.C., Coulter, J.P., Cavity pressure and part quality in the injection molding process The Science, Automation, and Control

of Material Process Involving Coupled Transport and Rheology Changes, vol. 89, *ASME*, 1999, pp. 7–17.

6. Stokes, V.J., Thermoplastics as engineering materials: the mechanics, materials, design, processing link, Trans. *ASME* 117 (1995) 448–455.

7. Gahleitner, M., Melt rheology of polyolefins, *Prog. Polym. Sci.* 26, (2001), 895–944.

8. Chum, P.S., Swogger, K.W., Olefin polymer technologies – History and recent progress at The Dow Chemical Company, *Progress in Polymer Science* 33 (2008) 797–819.

9. Vergnes, B., Vincent, M., Demay, Y., Coupez, T., Billon, N., and Agassant, J., Present challenges in the numerical modeling of polymer forming processes, *The Canadian Journal of Chem Eng.*, 80, 2002, 1143–1152.

10. Gahleitner, M., Wolfschwenger, J., Fiebig, J., Neissl, W., *Macromol. Symp.*, 2002, 185, 77–87.

11. Housmans, J.W., Gahleitner, M., Peters, G.W.M., Meijer, H.E.H., *Polymer* 2009, 50(10), 2304–2319.

12. Doak, K.W., *Encycl. Polym. Sci. Eng.*, 2nd edn., (ed., Kroschwitz, J.I.), Wiley Interscience, 1986, vol 6, 386.

13. Wolcott, M.P., Production methods and platforms for wood plastic composites. In: *Proceedings of the non-wood substitutes for solid wood products conference*, October 2003, Melbourne. 12 pp.

14. Dumitrescu, O.R., Baker, D.C., Foster, G.M., and Evans, K.E., *Polymer Testing* 24 (2005) 367–375.

15. Min, B.H., A study on quality monitoring of injection-molded parts, *J. Mater. Process. Technol.*, 136(2002) 1.

16. Sun, S.H., *Computers in Industry*, 55 (2004) 148.

17. Liang, J.Z., Ness, J.N., The calculation of cooling time in injection moulding, *Journal of Materials Processing Technology* 57 (1996) 62–64.

18. Hill, D., Further studies of the injection moulding process, *Appl. Math. Modelling* 1996, Vol. 20, p. 719–730.

19. Chen, C.S., Determination of the Injection Molding Process Parameters in Multicavity Injection mold, *Journal of Reinforced Plastics and Composites*, 2006; 25, 1367.

20. Cutkosky, M.R., Tenenbaum, J.M., *CAD/CAM Integration Through Concurrent Process and Product Design*, Longman. Eng. Ltd., 1987, p. 83.

21. Menges, G., Mohren, P., *How to Make Injection Molds*, second ed., Hanser Publishers, New York, 1993, p 129.

22. Brian, C., Sachin, P., Rebecca, H., Blaine, L., Jose, C., *Polymer Engineering and Science* (2006), 46(7), 844–852.

23. John Bozzwli, know-how injection molding, *Plastics Technology*, Feb., 2010, 2010, p. 13.

24. Tatara, R.A., Sievers, R.M., and Hierzer, V., *Journal of Materials Processing Technology*, 176 (2006) 200–204 2001.

25. Kwack, T.H., and Han, C.D., *J. Appl. Polym. Sci.*, 1983, vol 28, 3419.

26. Attalla, G., and Bertinotti, F.J., *Appl. Polym. Sci.*, 1983 , vol 28 , 3503.

27. Han, C.D., Kim, Y.J., Chung, H.K, and Kwack, T.H., *J. Appl. Polym. Sci.*, 1983, vol 28, 3435.
28. Kuhn, R., and Kromer H., *Colloid Polym. Sci.*, 1982, vol 260, 1083.
29. Barnes, H.A., Hutton, J.F., and Walters, K., *An introduction to rheology*, Elsevier, Oxford, U.K. 1989.
30. Pandelidis, I., Zou, Q., Optimization of injection molding design. Part I: Gate location optimization, *Polym. Eng. Sci.*, 30 (1990) 873–882.
31. Lee, B.H., Kim, B.H., Automated selection of gate location based on desired quality of injection molded part, SPE Annual Technical Conference, *ANTEC 1995*, pp. 554–560.
32. Menges, G., Bangert, H., Measurement of coefficients of static friction as a means of determinating opening and demoulding forces in injection moulds, *Kunstst. Ger. Plast.*, 71 (1981) 552.
33. Gao, F., Patterson, W.I., Kamal, M.R., Cavity pressure control during the cooling stage in thermoplastic injection molding, *Polym. Eng. Sci.*, 36 (1996) 2467.
34. Dininger, J. In *IEEE Industry Applications Society Annual Meeting* 1994; Vol. 3, 2159–2164.
35. Pierick, D., Noller, R. In *ANTEC '91*, 1991, pp. 252–258.
36. Gao, F., Patterson, W.I., Kamal, M.R., *Polym Eng Sci.*, 1996, 36(9), 1272–1285.
37. Huebner, K.H., Thornton, E.A., Byrom, T.G., *The Finite Element Method for Engineers*, 4th ed., Wisley, 2001, p. 1.
38. Zhao, C., and Gao, F., (1999). Melt temperature profile prediction for thermoplastic injection molding, *Polymer Engineering and Science*, 39(9), 1787–1801.
39. Chiang, K.T., and Chang, F.P., (2006). Application of grey-fuzzy logic on the optimal process design of an injection-molded part with a thin shell feature, *International Communications in Heat and Mass Transfer*, 33, 94–101.
40. Wu, C.H., and Liang, W.J. (2005). Effects of geometry and injection-molding parameters on weld-line strength. *Polymer Engineering and Science*, 45(7), 1021–1030.
41. Shen, Y.K., Liu, J.J., Chang, C.T. and Chiu, C.Y. (2002), "Comparison of the results for semisolid and plastic injection molding process", *Int. Comm. Heat Mass Transfer*, Vol. 29, pp. 97–105.
42. Pontes, A.J., Pouzada, A.S., Ejection force in tubular injection moldings. Part I: Effect of processing conditions, *Polym. Eng. Sci.* 44 (2004) 891.
43. Rheinfeld, D., Influencing moulding quality during injection molding, in: Welling, M.S., (translated), *Injection Moulding Technology*, VDI-Verlag BmbH, Düsseldorf, 1981.
44. Burke, C., Malloy, R., An experimental study of the ejection forces encountered during injection molding, *SPE ANTEC Tech. Pap.* 37 (1991).

45. Balsamo, R., Hayward, D., Malloy, R., An experimental evaluation of ejection forces: frictional effects, *SPE ANTEC Tech. Pap.* 39 (1993).
46. Lim L.T., Auras, R., Rubino, M., *Progress in Polymer Science*, Volume 33, Issue 8, 2008, Pages 820–852.
47. Kurtaran, H., and Erzurumlu, T., (2006). Efficient warpage optimization of thin shell plastic parts using response surface methodology and genetic algorithm. *International Journal of Advanced Manufacturing Technology*, 27(5/6), 468–472.
48. Mary E. Kinsella, Blaine Lilly, Benjamin E. Gardne, and Nick J. Jacobs, Experimental determination of friction coefficients between thermoplastics and rapid tooled injection mold materials, *Rapid Prototyping Journal*, Volume 11, Number 3, 2005, 167–173.
49. Pun, K.F., Hui, I.K., Lewis, W.G., Lau, H.C.W., A multiple-criteria environmental impact assessment for the plastic injection molding process: a methodology, *J. Cleaner Prod.*, 11 (2002) 41.
50. Chen, X., Lam, Y.C., Li, D.Q., Analysis of thermal residual stress in plastic injection molding, *J. Mater. Process. Technol.*, 101 (1999) 275.
51. Park, S,J., Kwon, T.H., Optimal cooling system design for the injection molding process, *Polymer Eng Sci.*, 1998;38(9):1450–62.
52. Dimla, D.E., Camilotto, M., and Miani, F., Journal of Materials Processing Technology 164–165 (2005) 1294–1300.
53. Reig, M.J., Segui, V.J., Zamanillo, J.D., Jose, D., *Journal of Polymer Engineering* (2005), 25(5), 435–457.
54. Duncan, R.E., and Zimmerman, A.B., Rotational molding of high density polyethylene powders, *Society of Plastics Engineers*, Regional Technical conference, p. 50, (march 1969).
55. Keller A., Kolnaar, H.W.H., *Flow induced orientation and structure formation*. In: Meijer HEH, editor. Processing of polymers, vol. 18. Weinheim, Germany: VCH; 1997.
56. Collins, C., *Assembly Automation*, 1999, 19(3), 197–202.
57. Becker, K.H., Lippmann, H., and Teubl, E., Plastic behaviour of metals and strain localization, *Naturwissenschaften*, 72 (1985) 633–639.
58. Gächter, R, Müller, H., *Plastics additives*, 3rd ed. Munich, Vienna, New York: Hanser Publishers; 1990.
59. Isayev, A.I., in: *Proceedings of the 23rd Israel Conference of Mechanical Engineering*, 1990, Paper #5.2.3, p. 1.
60. Seow, L.W., Lam, Y.C., *Journal of Materials Processing Technology*, 72 (1997) 333–341.
61. Pantani, R., Coccorullo, I., Speranza, V., Titomanlio, G., Modeling of morphology evolution in the injection molding process of thermoplastic polymers, *Prog. Polym. Sci.* 30 (2005) 1185–1222.
62. Wong, H.Y., Fung, K.T., and Gao, F., *Sensors and Actuators*, A 141 (2008) 712–722.

63. Blom, H., Yeh, R., Wojnarowski, R, and Ling, M., *Thermochimica Acta*, 442 (2006) 64–66.
64. Wang, K.K., Zhou, J., Sakurai, Y., In *ANTEC '99*, 1999; pp. 611–615.
65. Wong, C.M., Chen, C.H., Isayev, A.I., *Polymer Engineering and Science*, 30 (1990) 1574.
66. Bruker, I., Bulch, G. S., *Polym Eng Sci*, 1989, 29, 258–267.
67. Huang, M.C., Tai, C.C., The effective factors in the warpage problem of an injection-molded part with a thin shell feature, *J. Mater. Process. Technol.*, 110 (2001) 1–9.
68. Chen, Z.,Turng, L., A Review of Current Developments in Process and Quality Control for Injection Molding, *Advances in Polymer Technology*, Vol. 24, No. 3, 165–182 (2005).
69. Shapire, J., Hamlmes, A.L., Pearson, J.R.A., *Polym Eng Sci* 1976, 17, 905–918.
70. Breitenbach, J., Melt extrusion: from process to drug delivery technology, *Eur. J. Pharm. Biopharm.* 54 (2002) 107–117.
71. Hunk-Kuk, O., and Byung-Woo, R., Effects of process variables on the deformation field in extrusion through conical dies, *J. Mech. Technol.*, 11 (1984) 71–86.
72. Avitzur, B., *Handbook of Metal Forming*, Wiley, New York, 1983.
73. Isayev, A.I., Wong, C.M., Zeng, X., *SPE ANTEC Tech. Pap.*, vol. 33, 1987, p. 207.
74. Vormeulen, J.R., Gerson, P.M., *Chem Eng Sci* 1971, 26, 1445–1455.
75. Zweifel, H., *Stabilization of polymeric materials*, Berlin, Heidelberg, New York: Springer-Verlag; 1998.
76. Isayev, A.I., Wong, C.M., Zeng, Z., *Advanced Polymer Technology*, 10 (1990) 31.
77. Levin, V.Y., Kim, S.H., Isayev, A.I., *Rubber Chemical Technology*, 69 (1995) 104.
78. Isayev, A.I., Yushanov, S.P., J. Chen, *Journal of Applied Polymer Science*, 59 (1996) 803.
79. Ivanov, A.V., Bilalov, Y.M., Ismailov, T.M., *Prikl. Reol. Techenie Dispersnykh Sist. (RUSS)*, (1981) 57, From CA, 98 (18) (1983) 1447–96.
80. Keishiro Oda., *JPN Patent*, 91 253, 323, 1991, From CA, 116 (12) (1992) 108–130.
81. Aldea, C., Marikunte, S., and Shah, S.P., *Adv. Cem. Based Mater.*, 8 (1998) 47.
82. Chung, C. I., *Extrusion of Polymers: Theory & Practice*, Hanser Gardner, Cincinnati (2000).
83. Dealy, J.M., and Wissbrun, K.F., *Melt Rheology & Its Role in Plastics Processing: Theory & Applications*, Chapman & Hall, New York (1995).
84. Baird, D.G., and Collias, D.I., *Polymer Processing: Principles & Design*, John Wiley & Sons, New York (1998).
85. Krishnaswamy, R.K., Rohlfing, D.C., Sukhadia, A. M., and Slusarz, K.R., *Polymer Engineering and Science*, 44, 2266 (2004).

86. Walsh, E.B., Gallucci, R.R., and Courson, R.,(1991) "High Density Thermoplastic Polyesters," *ANTEC '91*. Lancaster, PA: Technomic Publishing Co., Inc., p. 1334.

87. Mudalamane, R., Bigio, D.I., Tomayko, D.C., and Meissel, M., Behavior of Fully Filled Regions in an Non-Intermeshing Twin Screw Extruder," Poly. Eng. & Sci., 43, 8, 1466–1476, 2003.

88. Rajath Mudalamane and David I. Bigio, *AIChE Journal*, Dec., 2003 Vol. 49, No. 12, p. 3150–3160.

89. Carneiro, O.S., Caldeira, G., Covas, J.A., Flow patterns in twin-screw extruders, *Journal of Materials Processing Technology*, 92–93 (1999) 309–315.

90. Janssen, L.P.B.M., *Twin Screw Extrusion*, Elsevier, Amsterdam (1978).

91. White, J.L., Intermeshing Counter-Rotating Twin Screw Extrusion Technology, *Twin Screw Extrusion*, Hanser Verlag, Berlin (1991).

92. Shon, E., Chang, D., and White, J.L.,"A Comparative Study of Residence Time Distributions in a Kneader, Continuous Mixer, and Modular Intermeshing Co-Rotating and Counter-Rotating Twin Screw Extruders," *Int. Polymer Processing*, 14, 44 (1999).

93. Raut, J.S., Naik, V.M., and Jongen, T.R., Efficient Simulation of Time-Dependent Flows: Application to a Twin Screw Extruder, *AIChE Journal*, Aug., 2003 Vol. 49, No. 8, 1933–45.

94. Levenspiel, O., *Chemical Reaction Engineering*, Wiley, New York, 1972.

95. Weiss, R.A., and Stamato, H., "Development of an Ionomer Tracer for Extruder Residence Time Distribution Experiments," *Poly. Eng. Sci.*, 29, 134, 1989.

96. Cassagnau, P., Mijanggos, C., and Michel, A., "An Ultraviolet Method for the Determination of the Residence Time Distribution in a Twin Screw Extruder," *Poly. Eng. Sci.*, 31, 772, 1991.

97. Kim, P.J., and White, J.L., "On-Line Measurement of Residence Time Distribution a Twin Screw Extruder," *Int. Poly. Process.*, IX, 2, 108 1994.

98. Chen, T., Patterson, W.I., and Dealy, J.M., "On-Line Measurement of Residence Time Distribution in a Twin Screw Extruder," *Int. Poly. Process.*, X, 1, 3, 1995.

99. Wetzel, M.D., Shih, C.K., and Sundararaj, U., "Determination of Residence Time Distribution During Twin Screw Extrusion of Model Fluids," *SPE ANTEC Tech. Papers*, 3, 3707, 1997.

100. Denn, M., Extrusion instabilities and wall slip, *Annu. Rev. Fluid Mech.*, 33 (2001) 265.

101. Crawford, R.J., (2002) *Plastics Engineering*, 3rd edn, Butterworth-Heinemann Publisher, Oxford, England.

102. Rao, N., and O'Brien, K., (1998). *Design Data for Plastics Engineers*, 1st edn, Hanser/Gardner Publications, Inc., Cincinnati, USA.

103. Brown, E.C., Kelly, A.L. and Coates, P.D. (2003). Effect of Extrusion Die Geometry on Molecular Orientation of Unfilled Polyethylene, In: *ANTEC 2003*, pp. 1464–1468, Nashville, USA.

104. Revesz, H., Hubeny, H., Continuous measurement and control of viscosity through the extrusion process. Proc. 1, *3rd IFAC Conference on Instrumentation and Automation*. In: The Paper, Rubber and Plastics Industries, Brussels, 1976, p. 69.

105. Paschke, E., *Kirk-Othmer Encycl. Chem. Technol.*, Vol. 16, 3rd edn., (eds. Grayson, M., and Eckroth, D.,), Wiley Interscience, 1981, 440.

106. Lai, E., Yu, D.W. Polym Eng Sci 2000, 40, 1074–1084.

107. Dormeier. S., *Extruder control. 4th IFAC Conference* in: The Paper, Rubber, Plastics and Polymerization Industries, Ghent, Belgium, 1980, p. 551.

108. Lindt, J.T., "Mathematical modelling of melting of polymers in a single-screw extruder; a critical review", *Polymer Engineering and Science*, 25, 10 (1985) 585.

109. Curry, J.E., Kinni, A. In Proceedings of the 52nd SPE ANTEC, San Francisco, CA, 1994; pp. 172–177.

110. Yung, K.L., Xu, Y., Lau, K.H., Transient melting models for the three stages of reciprocating extrusion, *Journal of Materials Processing Technology*, 139 (2003) 170–177.

111. Rauwendaal, C., Conveying and melting in screw extruders with axial movement, *Int. Polym. Process.*, 7 (1) (1992) 26–31.

112. Mitera, J., Michal, J., Kubat, J., and Kubelka, V.Z, *Anal. Chem.*, 1976, vol 281, 23.

113. MacSporran, W.C., *Bull. Brit. Soc. Rheol.*, 1, 5, 1959.

114. Goren, S.L., Middleman, S., and Gavis, J., *J. Appl. Polym. Sci*, 3, 367, 1960.

115. Bagley, E.B., Storey, S.H., and West, D.C., *J. Appl. Polym. Sci.* 7, 1661, 1963.

116. Schreiber, H.P., and Bagley, E.B., *Polym. Lett.* 1, 365, 1963.

117. Batchelor, J., and Horsfall, F., Die Swell in Elastic and Viscous Fluids, *Rubber and Plastics Research Association of Great Britain* 189, 1971.

118. Todd, D.B. In *Proceedings of the 50th SPE ANTEC*, Detroit, MI, 1992; pp. 2528–2533.

119. Chan Hua, T. In Proceedings of the *53rd SPE ANTEC*, Boston, MA, 1995; pp. 302–307.

120. Kalika, D.S. , and Denn, M.M., *J. Rheol*, 31, 815 (1987).

121. Beynon, D.L.T., and Glyde, B.S., *Brit. Plast.*, 33, 414, 1960.

122. Graessley, W.W., Glasscock, S.D. and Crawley, R.L., *Trans. Soc. Rheol.*, 14, 519, 1970.

123. Han, C.D., *Rheology in polymer processing*, Academic, New York, 1976, Chap. 5.

124. Weeks, D.J. and Allen, W. J., Screw Extrusion of Plastics, *J. mech. Engng Sci.* 4, (4) 380–399 (1962).

125. Rudin, A., and Chang, R.J., *J. Appl. Polym. Sci*, 22, 781, 1978.
126. Nielsen, L.E., *Polymer Rheology*, Dekker, New York, 1977, chap. 7.
127. Shroff, R., and Shida, M., Effect of molecular weight and molecular weight distribution on elasticity of polymer melts, *ANTEC* (1977).
128. Tadmor, Z., and Gogos, C.G., *Principles of Polymer Processing*, Interscience, New York, 1979, Chap 15.
129. Jim Frankland, *Plastics Technology*, 55, Nov. 2009, p. 21.
130. Kalyon, L.M., Lawal, A., Yazici, R., Yaras, P., and Railkar, S., Modeling and Studies of Twin-Screw Extrusion of Filled Polymers, Polymer Engineering Science, June 1999, Vol. 39, No. 6, 1139–1151.
131. Yang, B., and Lee, L.J., "Process control of profile extrusion using thermal method. Part:I, Mathematical modelling and systems analysis", *Polymer Engineering and Science* 28,11 (1988) 697–707.
132. Pabedinskas, A., Cluett, W.R., Balke, S.T., *Polymer Engineering and Science* 31 (1991) 365.
133. Somers, S.A., Spalding, M.A, Dooley, J., Hyun, K.S., *Proc. SPE ANTEC*, Paper #659, San Francisco (2002).
134. Van Zuilichem, D.J., Kuiper, E., Stolp, W., Jager, T., Mixing effects of constituting elements of mixing screws in single and twin screw extruders, *Powder Technology* 106 (1999) 147–159.
135. Heinze, M., *Surface and Coatings Technology*, 105 (1998) 38–44.
136. Elbirli, B., Lindt, J.T., Gottgetreu, S.R., Baba, S.M., *Polym Eng Sci.*, 1984, 24, 988–999.
137. Cox, A.P.D., Fenner, R.T., *Polym Eng Sci*, 1980, 20, 562–571.
138. Tadmor, Z., Lipshitz, S.D., and Lavie, R., Dynamical model of a plasticating extruder. *Polymer Engng Sci.*,14(2), 112–119(1974).
139. Gao, D.M., Nguyen, K.T., Hetu, J.F., Laroche, D. and Garcia-Rejon, A., (1998), "Modeling of industrial polymer processes: injection molding and blow molding", *Advanced Performance Materials*, Vol. 5 No. 1–2, pp. 43–64.
140. Zahavich, A.T.P., Latto, B., Takacs, E., Vlachopolulos, J., The effect of multiple extrusion passes during recycling of high density polyethylene, *Polym Techn.*, 16:11–24, 1997.
141. Huang, H.X., Lia, Y.Z., and Denga,Y.H., *Polymer Testing*, 25 (2006) 839–845.
142. Menges, G., Esser, K., Hüsgen, U., and Kunze, B., Process Optimization in Stretch Blow molding, *Advances in Polymer Technology*, 6(3), 389–397, 1986.
143. Attara, A., Bhuiyanb, N., Thomson, V., Manufacturing in blow molding: Time reduction and part quality improvement, *Journal of materials processing technology*, 204 (2008) 284–289.
144. Shelby, M.D., "Effects of Infrared Lamp Temperature and Other Variables on the Reheat Rate of PET," *SPE ANTEC Tech. Papers* 37 1420–1424 (1991).

145. Diraddo, R.W., Garcia-Rejon, A., "Experimental and Theoretical Investigation of the Inflation of Variable Thickness Parisons," Polym. Eng. Sci., 34, 13, 1080–1089, 1994.
146. Erbulut, D.U., Masood, S.H., Tran, V.N., Sbarski, I., A Novel Approach of Measuring the Dielectric Properties of PET Preforms for Stretch Blow Moulding, *Journal of Applied Polymer Science*, Vol. 109, 3196–3203 (2008).
147. Haihong Xu, David O. Kazmer, Thermoforming Shrinkage Prediction, *Polymer Engg. and Sci*, 2001, 41, 9.
148. Bhattacharyya, D., Bowis, M., Jayaraman, K., Thermoforming woodfibre – polypropylene composite sheets, *Composites Science and Technology*, 63 (2003) 353–365.
149. Cogswell, F.N., *Thermoplastic aromatic composites*, Oxford: Butterworth, 1992.
150. Schuster, J., Friedrich, K., Modeling of the mechanical properties of discontinuous-aligned-fiber composites after thermoforming, *Comp Sci Tech.*, 1997;57:405–13.
151. Dharia, Amit., *Annual Technical Conference – Society of Plastics Engineers* (2006), 64th 2605–2609.
152. Liu, S.C., Hu, S.J., and Woo, T.C., Transactions of the ASME, *Journal of Mechanical Design*, 118, 62–67. 1996.
153. Fathallall, A.K., and Dixon, J.R., *American society of Mechanical Engineers*, Design Engineering Devision [Publication] DE, pp. 9–17, Minneapolis 1994.
154. Gruenwald, G. *Thermoforming; A Plastics Processing Guide*, Lancaster, PA: Technomic Publishing Co. (1987).
155. Irwin, D., *Modern Plastics Encyclopedia*, 1986–1987 New York: McGraw Hill Inc., pp. 322–330 (1986).
156. Hartley, P., Pillinger, I., and Sturgess, C., *Numerical Modelling of Material Deformation Processes*, Springer, Berlin, 1992.
157. Bellehumeur, C., Bisaria, M.K., Vlachopoulos, J. Polym. Eng. Sci. 1995, 40, 1973.
158. Rao, M.A., Throne, J.L., *Polym. Eng. Sci.* 1972, 12, 237.
159. Throne, J.L., *Polym. Eng. Sci.* 1972, 12, 335.
160. Howard, H.R., Variables in rotomolding that are controllable by the molder, *Association of Rotational molders*, Oct 1977.
161. Greco, A., and Maffezzoli, A., (2004). Powder-Shape Analysis and Sintering Behavior of High-Density Polyethylene Powders for Rotational Molding, *Journal of Applied Polymer Science*, 92(1): 449–460.
162. Oliveira, M.J., Cramez, M.C., Crawford, R.J., Structure – properties relationship in rotationally moulded polyethylene. *J Mater Sci* 1996; 31(9): 2227–40.
163. MacAdams, J., "How to Predict Physical Properties of Rotomoulded Parts", *SPE (Society of Plastics Eng.) Reg. Tech. Conf.*, Chicago, USA, Oct. (1975) p. 64.

164. Pick, L.T., Jones, E.H., *Polymer Engg. and Sci.*, 4, 43, 2003.
165. Ramazzotti, D.J., *Rotational molding – the state of the art*, The Society of Plastic Engineers, Regional technical conference (Oct 1975).
166. Crawford, R.J., *Progress in rubber and plastic technology*, 6, 1–29, (1991).
167. Spence, A.G., and Crawford, R.J., The effect of processing variables on the formation and removal of bubbles in Rotationally molded product, *Polym. Eng. Sci.*, 1996, 36, 7, 993–1009.
168. Crawford, R.J., and Nugent, P.J., *Plastics and Rubber processing and application*, 17, 23 (1992).
169. Maier, R.D., *Kunststoffe Plast Europe*, 39, 45, 1999.
170. Throne, J.L., Polym Eng Sci., 1976, 16, 257.
171. Throne, J.L., Sohn, M.S., *Adv. Polym. Techn.*, 1989, 9, 281.
172. Bisaria, M.K., Vlachopoulos, M, *J. Int. Polym Proc.*, 1997, 12, 165.
173. Cramez, M.C., Oliveira, M.J., Crawford, R.J., Effect of cooling rate and nucleating agents on a rotational moulding grade of polypropylene. *J Mater Sci* 2001;36(9):2151–61.
174. Liu, S.J., Lai, C.C., Lin, S.T., *Polym Eng Sci* 2000, 40, 473.
175. Bisaria, M.K., Takács, E., Bellehumeur, C.T., and Vlachopoulos, J., *Rotation*, 3(4), 12, 1994.
176. Bell, G.L., *Rotational Molding: Design, Materials, Tooling and processing*, Hanser/Gardner Publications, Inc., Cincinnati, 1998.
177. Crawford, R.J., *Rotational Moulding of plastics*, second edition, John Wiley and sons Inc., 1996.
178. Throne, J.L., *Thermoplastic foams*, Sherwood Technologies, Inc., Sherwood Publishers, Ohio 1996.
179. Crawford, R.J., and Throne, J.L., *Rotational molding technology*, William Andrew Publishing, NY, 2002.
180. Crawford, R.J., Nugent, P.J., *Plas. Rubb. Proc. Appl.*, 1989, 11, 107.
181. Nugent, P.J., Crawford, R.J., Xu, L., *Adv. Polym. Techn.*, 1992, 11, 181.
182. Crawford, R.J., *3rd IDFC Conference Proceedings*, Galway, Ireland (1985).
183. Crawford, R.J., *Journal of material processing technology*, 56 (1996) 263–271.
184. Liu, S.J., *Int. Polym Proc.* 1998, 13, 88.
185. Bellehumeur, Vlachopoulos, *SPE ANTEC technical papers*, 44, 1112 (1998).
186. Wigotsky, V., *Plastics Engineering*, Feb 1998, pp 18–23.
187. Nugent, P.J., and Crawford, R.J., Rotational molding of plastics, R.J. Crawford, ed., John Wiley and Sons Inc., New York 1992.
188. Throne, J.L., *Plastics process engineering*, pp. 579–614, Marcel Dekker, New York 1979.
189. Bisaria, M.K., Takács, E., Bellehumeur, C.T., and Vlachopoulos, J., *Plastics Engineering*, Feb 1998, pp. 29–31.
190. Pop-Iliev, R., Xu, D., and Park, C.B. (2004). Manufacturability of Fine-Celled Cellular Structures in Rotational Foam Molding, *Journal of Cellular Plastics*, 40(1): 13–25.

191. Briken, F., and Potente, H., (1980), *Polym. Eng. Sci.*, 20: 1009.
192. Malpass, V.E., Kempthorn, J.T. and Dean, A.F., *Plast. Eng.*, Jan 27 1989.
193. Muzzy, J.D., Wu, X. and Colton, J.S., (1990), *Polym. Comp.*, 11: 280.
194. Machida, T., and Lee, D., (1988), *Polym. Eng. Sci.*, 28: 405.
195. Yang, C., and Hung, S., Modeling and Optimization of a Plastic Thermoforming Process, *Journal of Reinforced Plastics and Composites*, 2004; 23; 109.

5

Troubleshooting – Problems and Solutions

In plastics processing, there are many factors that affect not only processing but also the end products. Different plastic materials may have various processing conditions, but they may need to match the change in raw material during processing. It is important to focus on the root causes of defects to identify and solve the problem that occurs during processing.

Observation is the key to solve any problem. The relations between the process and end products are answered by raw material and processing parameters used. The problems are interconnected with each other. Troubleshooting is a measured set consisting of powerful ideas and methods that will be useful in processing of plastics.

Troubleshooting is an attempt that combines science and technology to eliminate the problems and improve quality. The purpose of troubleshooting is to increase the process efficiency with quality of the end production. It is to identify, treat and evaluate to develop and maintain complex systems where the symptom of a problem can have many causes. It begins with description of the background of the problem and the situation to be analyzed.

Troubleshooting is corrective actions starting from a correct understanding of the facts to process improvements. Troubleshooting will decide the boundaries of the problem and is one of the most challenging tasks in the industry. It is achieved by minimizing the defects. The relations between the process and end products are answered by raw material and processing parameters used.

In plastics processing, the problems usually occur in one of the following areas:

- Raw material
- Additives involved in the formulation
- Processing equipment in operation
- Process control and their settings

Some of the problems can have several different potential causes in troubleshooting. Hence troubleshooting is

- To identify and recognize the problem areas with respect to type, level and status.
- To analyze the problems
- To approach the problems.
- To solve the problems in a practical and concrete way.

Troubleshooting is a guideline to establish the information and data collection to make comprehensive decisions concerning the problems and defects. It is useful for making improvements in individual situations. Troubleshooting is a very powerful way to eradicate the barriers that exist within various areas of processing. Troubleshooting techniques can be used for two main purposes

1. Discover problem areas in the processing
2. Solve the problem in a practical and concrete way.

The success of troubleshooting is based on

- Understand the correct problem situation
- Identify the problems clearly and
- Appropriate solution to be applied to solve the problem.

5.1   Troubleshooting – Requirements

Success of troubleshooting is to understand the correct situation of the problems and solution to achieve and maintain processing operation.

The problem solving approach can be attained by

- Plan, update, and experience in the processing.
- Recording of processing condition
- Control or condition to be changed only one at a time
- Sufficient time to be given to take the change
- Log of each change has to be maintained
- Narrow the range of areas of which the problem belongs. It may be related to machine, molds, or dies, operating controls, material, part design, humidity, etc.,

Requirements for efficient troubleshooting

- Understanding the process, condition of the equipment, collection of historical data, and information on the feedstock
- Temperature measuring device, data acquisition systems

Upsets versus development problems require

- Troubleshooting techniques
- Systematic approach.

5.2   Injection Molding – Troubleshooting

With exceptional physical and mechanical properties, the injection molding of thermoplastic material has become an important process in industry. There are still some problems that confound the overall success of the technology [1].

Defects are an important limitation in terms of quality and productivity in most plastics processing technique. Defects and instabilities occur during processing of plastics. Defects can be classified into: sink marks, streaks, gloss differences, visible weld lines, jetting, diesel effect (burns), record grooves effect, stress whitening or cracking, incompletely filled parts, flash, visible ejector marks, deformation during demolding, flaking of the surface, cold slugs or cold flow lines, entrapped air and blister formation, dark spots, and dull spots near the sprue.

Defects often become weak spots for fracture. Mechanical strength is directly correlated with defects. The change in parameters has its own influences on other parameters. Hence process control becomes very complex. It not only affects the appearance but in many circumstances affects both its functionality and performance during its usable life. The feedstock rheology governed by the plastic material may also influence the occurrence of defects.

Defects of the products, such as warpage, shrinkage, sink marks, and residual stress, are caused by many factors during the production process. These defects influence the quality and accuracy of the products.

The defects are two types [2–4] in injection molding.

1. Molding defects. Figure 1 indicates molding defects during injection molding
2. Molded part defects. Figure 2 indicates the molded part defects occur during injection molding.

In troubleshooting of molded part defects, it is important to look at the shot size, injection speed, injection pressure, cushion, decompression, and nozzle tip. The following troubleshooting flow charts will help to resolve a majority of the problems. The forming of weld lines wherever polymer flow fronts meet is one of them [1].

Molded part defects are the defects identified after the product is produced. Problem of jetting occurs as shown in Figure 3.

The mold filling stage problem as a problem of combined energy and transport of momentum and applied to the transient and non-isothermal flow problem depends on many

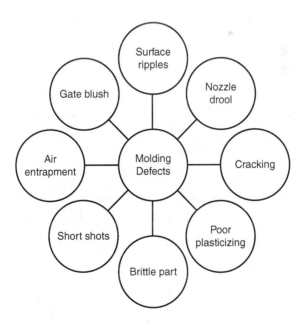

Figure 1. Molding defects.

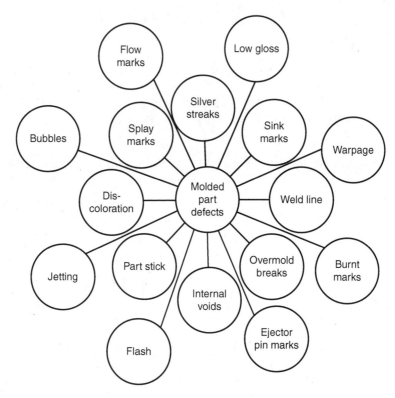

Figure 2. Molded part defects.

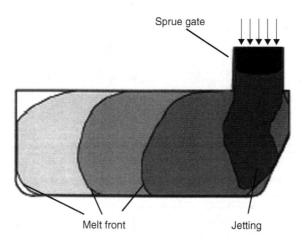

Figure 3. Schematic profile of melt front during mold filling. (Reproduced with permission from S. Krug et al./Journal of the European Ceramic Society 22 (2002) 173–181 © 2001 Elsevier Science Ltd. All rights reserved.) Ref. [5].

factors connected with processed plastic, injection mold, injection machine, and process conditions.

Problem of flow is as shown on a molded surface in Figure 4.

Sharkskin is the first limiting factor affecting the production rates. It is basically a surface defect and an exit phenomenon. It is characterized by a small scale and high frequency roughness. [7–13].

Degraded materials tend to stick, especially at the nozzle; temperature needs to be controlled tightly. Crazing of the part surface can also be a sign of the cause of sticking. A sprue that is too soft or not frozen can cause sticking. Scratches, unbalanced filling, over packing, polishing, and sprue created during use can cause sticking of the sprue or part in the mold. Over packing is the main cause related to sprue sticking. Another possible cause is plate-out on the mold; excessive shrinkage, mold, or material temperatures can also be a source of sticking in the mold.

Undercuts removal from the mold will solve the sticking problem. In case of mold undercuts insufficient draft has to be

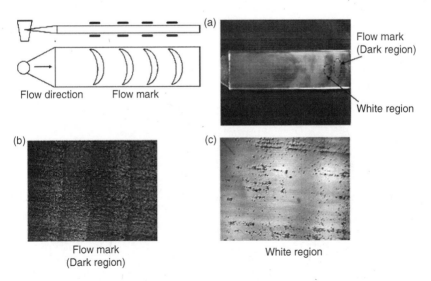

Figure 4. Flow mark of mold part surface. a. Typical photograph of the molded part surface b. flow mark (Dark region) C. white region (Reproduced from Chao-Chyun An, Ren-Haw Chen, journal of materials processing technology 201 (2007) 706–709.© 2007 Elsevier B.V. All rights reserved). Ref.[6].

modified in some cases to resolve many of the mold related problems.

5.2.1   Part Sticking in Cavities

Material sticking will reduce the production efficiency. Too high polished mold can cause vacuum to form during molding that holds the part to the mold and may not allow good part to release. In such case, air pipette can be used to release the molded vacuum holding part from the mold. It is common for ribs to have a poor texture on them, which will cause the part to stick in the mold. Sticking ribs in a mold will often give the appearance of a sink, but it is a pull.

Flow chart 5.2.1 illustrates the schematic approach to solve the problem of sticking in cavities.

Problems and Troubleshooting – Injection Molding

Flow chart 5.2.1 – Part sticking in cavities

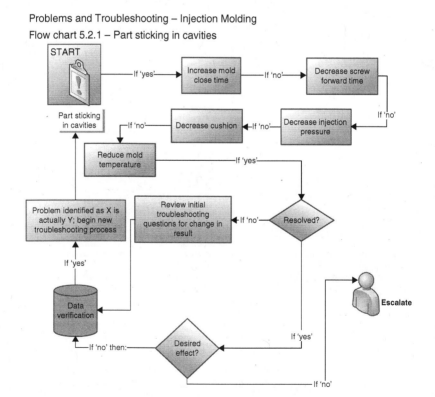

5.2.2   Part Sticking in Core

Flow chart 5.2.2 illustrates the schematic approach to solve the problem of sticking in core. Part sticking in core is due to frictional forces occurring during injection pressure.

Problems and Troubleshooting – Injection Molding

Flow chart 5.2.2 – Part sticking in core

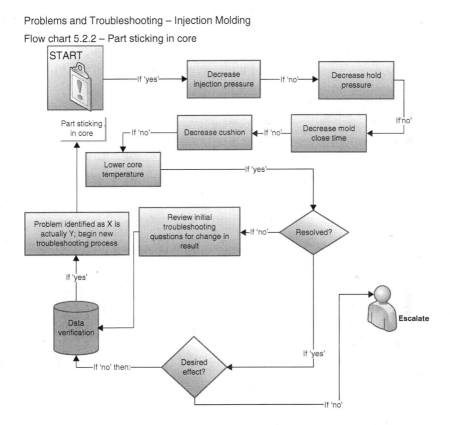

5.2.3   Discoloration

Discoloration is a serious commercial problem in polymers [14–15]. The two contributors are: (a) impurities with non-homogeneity presents in polymer from the manufacturing which absorb UV-VIS light [16] and (b) sacrificed or depleted consumption of stabilizers arises during life time as a consequence of reactions of stabilizers with peroxide radicals, oxidizing metallic impurities, and pollutants from atmosphere and/or with catalysts residue present in the polymer during polymerization [17]. Discoloration can be monitored by changes in yellow index.

Flow chart 5.2.3 illustrates the schematic approach to solve the problem of discoloration.

Examination is required in the case of nozzle and cylinder for hold up points. Feed contamination has to be eliminated and no stoppage in the check ring will help to eliminate the problem. Finally purge the machine with the purging compound.

Problems and Troubleshooting – Injection Molding

Flow chart 5.2.3 – Discoloration

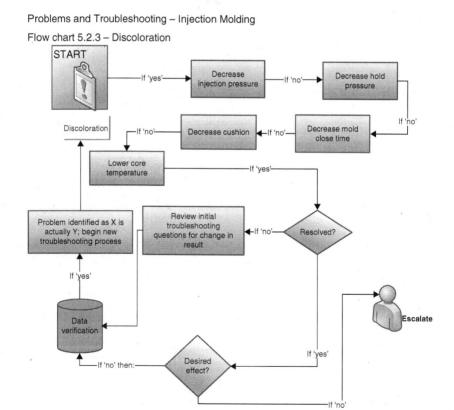

5.2.4   Burnt Marks (Diesel Effect)

Problems and Troubleshooting – Injection Molding

Flow chart 5.2.4 – Burnt marks (Diesel effect)

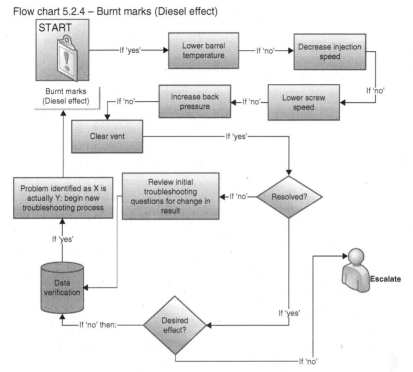

Flow chart 5.2.4 illustrates the schematic approach to solve the problem of burnt marks, also known as diesel effect. Overstay of the material in the barrel leads to the burnt marks. Machine size has to be considered to resolve the problem. There is a possibility of a vent getting clogged. Improve venting and in some cases the vent may be deepened. Increase in gate size may also be required in some cases.

5.2.5   Part Sticks During Ejection

Increase in overall cycle time, number of ejector pins, and increase in pin diameter on larger components may resolve the problem. Considering the use of pneumatic air ejection may also resolve the problem. Flow chart 5.2.5 shows the way to solve the problem of part sticks during ejection.

Problems and Troubleshooting – Injection Molding

Flow chart 5.2.5 – Part sticks during ejection

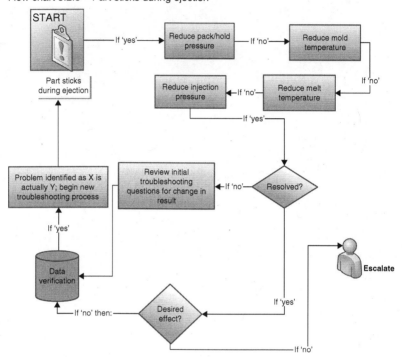

5.2.6   Jetting

Relocate gate so that the melt impinges off wall as it enters cavity. Flow chart 5.2.6 illustrates a schematic approach to the problem of jetting.

5.2.7   Dimensions Out of Specification

In such problem, tool dimensions have to be checked, before starting troubleshooting. Flow chart 5.2.7 illustrates the approach to solving the problem of dimensions out of specification.

Problems and Troubleshooting – Injection Molding
Flow chart 5.2.6 – Jettings

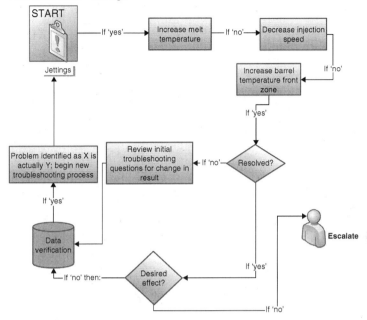

Problems and Troubleshooting – Injection Molding
Flow chart 5.2.7 – Dimensions out of specification

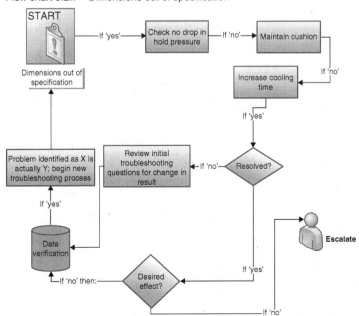

5.2.8 Bubbles

Bubbles can be seen in a plastic part. Insufficient or improper venting could cause the build up of gas resulting in bubble. High melt temperature could cause material to degrade and would cause bubbles. High mold temperature either through the whole mold or locally could cause bubbles and it needs to be monitored closely. In some cases gate size has to be increased to resolve the problem.

Problems and Troubleshooting – Injection Molding
Flow chart 5.2.8 – Bubbles

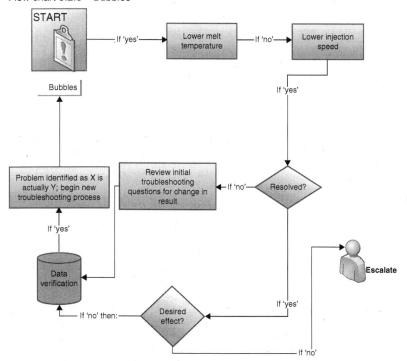

Flow chart 5.2.8 illustrates the schematic approach to the problem of bubbles.

5.2.9 Internal Voids

Voids are very similar to bubbles, but less noticeable. High melt temperature can cause degradation and cause voids to appear. Insufficient packing sometimes causes voids. It happens because the frozen layer of material stays against the tool while the part sinks from the inside out. Finally, either relocate or redesign the gate or increase runner; sprue may resolve the problem. Deepening vents or addition of vents helps the entrapped gas to escape. Flow chart 5.2.9 shows the schematic approach to the problem of internal voids.

Problems and Troubleshooting – Injection Molding

Flow chart 5.2.9 – Internal voids

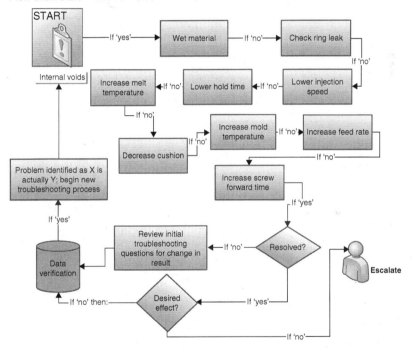

Flow chart 5.2.9 illustrates the schematic approach to the problem of internal voids.

5.2.10   Flash (Over Substrate or on Periphery of Part)

Flash can often be easily remedied. The parting line is free of any debris and properly maintained. Excess shot size or pack can often cause flash around the gate. Flash is a common part problem in injection molding. The flashing may be simply not enough clamping force. Flash stems from a mold or clamp alignment issue. These considerations have led to quality control problems, since the items are prone to cracking [18]. Insufficient clamp tonnage or pressure can result in poor shut off and will cause the part to flash. Material that becomes watery with high melt temperature could cause to flash more easily.

Problems and Troubleshooting – Injection Molding

Flow chart 5.2.10 – Flash

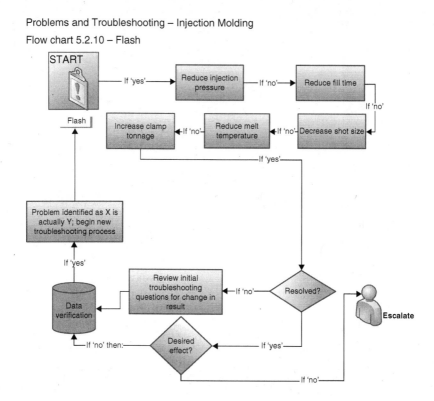

Mold to be properly fit in the machine and to be checked for substrate sinks and substrate support. Cut the tool to obtain complete shut off with interference into substrate. Thickness of the vents has to be checked and in some cases this has to be reduced to avoid the flash. Flow chart 5.2.10 shows the way to solve problem of flash over substrate or on periphery of part.

Flash is a common part problem with a host of possible answers. If you have shorts and flash, you could have a mold or clamping alignment problem. This may not be the case if you are running a high speed, thin wall product – the problem may be simply not enough clamp force. To help establish whether flash stems from a mold or clamp alignment issue, check parting line mating.

5.2.11   Poor Weld Line

Two material fronts that come together and melt together to become one is called weld line. Two melt fronts are not at a high enough temperature to cause the weld line to be weak. Insufficient pressure applied can sometimes cause the weld line to be poorly welded together. Poor venting near the weld line can cause the material fronts to encounter a restriction that could cause a poor weld.

Problems and Troubleshooting – Injection Molding

Flow chart 5.2.11 – Poor weld line

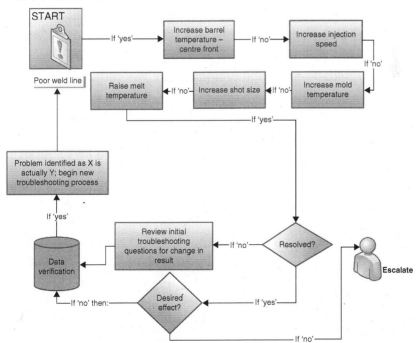

Weld lines occur when part or tool design features – like holes or ribs – split the melt flow and the separate flows do not weld completely when they meet. Weak weld lines occur when there is little or no polymer chain entanglement across the flow front boundaries.

One of the problems of injection molding process in most of the multi-gate mold designs is weld line or knit line which is formed when two or more separate melt fronts traveling in different directions meet. Weld line is undesirable when the strength, long term durability, and surface quality are a concern. It is observed due to multi-gate molding, existence of pins, inserts or cores within the mold cavity. Variable wall thickness and jetting may lead to weld line because of a sudden change in geometry along the flow path [19–24]. Adding glass fibers also diminishes weld line strength. Weak weld lines have their origins in material choice, part design, tooling, and processing.

Weld or knit lines are perhaps the most common and difficult injection molding defect to eliminate. They occur when melt flow fronts collide in a mold cavity. Material characteristics can affect the knitting of the melt fracture. A poor knit line can cause only cosmetic blemishes or it can significantly weaken the structural integrity of part strength.

In some cases, relocation of gate makes the flow pattern of the melt alter. Modify the vent location also in the cavity at the area of weld line and provide overflow well. Machine size has also to be considered to resolve the problem. Flow chart 5.2.11 illustrates the ways to solve the problem of poor weld line.

5.2.12   Low Gloss/Gloss Difference

Gloss is important to make a careful evaluation of the situation. Temperature or pressure too low will not allow the area to be packed out properly and can cause low or difference in gloss at two melt fronts. Improper vents around parting lines, on ejector pins, and on lifters can cause gas to come to the surface of the part and will give it a gloss difference.

Material traveled a long distance will sometimes not be packed out properly far from the gate. Larger gates or changing the gate location can help to pack that area better. Optimizing the hold on pressure may increase the gloss of the molded parts. Flow chart 5.2.12 shows the way to solve the problem of low gloss/gloss difference.

Problems and Troubleshooting – Injection Molding
Flow chart 5.2.12 – Low gloss/gloss difference

5.2.13   Overmold Breaks/Impinges Through Hollow Substrate

Either relocating the gate to the thickest section or avoiding gating to the thinnest wall area of the substrate may resolve the problem. Mold parts should be fully supported and substrate may need to be changed to resolve. Flow chart 5.2.13 shows a schematic approach to the problem of overmold breaks/impinges through hollow substrate.

Problems and Troubleshooting – Injection Molding

Flow chart 5.2.13 – Overmold breaks/impinges through hollow substrate

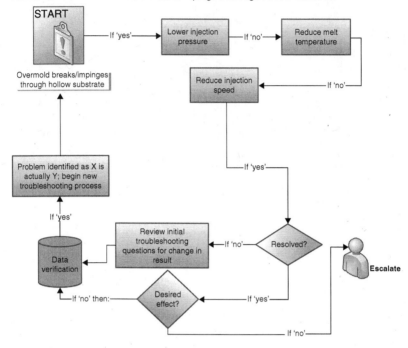

5.2.14 Warpage or Warped Parts

Optimizing or equalizing the mold temperature and uniformity of ejection or handling of parts after ejection from mold will help to resolve the problem. In case of counteract warp, trying differential mold temperatures may resolve the problem. Increase or design the substrate stiffness by the addition of glass reinforcement or thickness of ribs on substrate part structure. Flow chart 5.2.14 shows the schematic approach to the problem of warpage or warped parts.

5.2.15 Splay Marks/Delamination

Splay is caused by moisture and shear. The shear splay is also known as silver streaks. Splay caused by moisture is related to raw material and the shear splay, i.e., silver streaks are caused by processing. In case of splay, proper drying of the

Problems and Troubleshooting – Injection Molding

Flow chart 5.2.14 – Warpage or warped parts

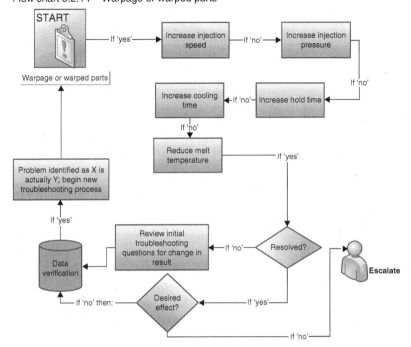

material is required. Some additives may be cooked out of the material. For example, polypropylene does not absorb moisture, but the talc filled will absorb. Hence, it is important to drive off that moisture. Flow chart 5.2.15 shows the ways to approach the problem of splay marks/delamination.

5.2.16   Flow Marks, Folds and Back Fills

Reposition or reposition to balance the flow to a thick section in the gate or reducing the runner diameter may resolve the problem. Surface texture can be added to part design and steel wall cavities also solve the problem by masking the flow marks. Flow chart 5.2.16 shows the schematic ways to solve the problem of flow marks, folds, and back fills. High injection speed will improve the flow length and will also reduce the visibility of flow marks.

Problems and Troubleshooting – Injection Molding

Flow chart 5.2.15 – Splay marks/Delamination

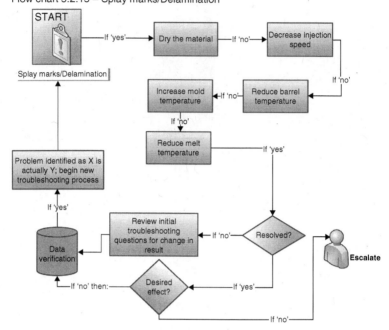

Problems and Troubleshooting – Injection Molding

Flow chart 5.2.16 – Flow marks, folds and back fills

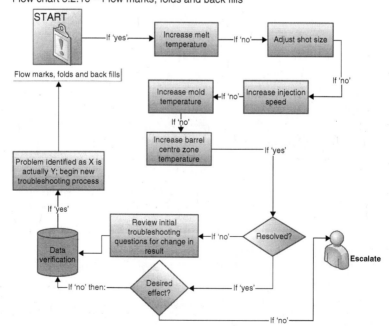

5.2.17   Ejector Pin Marks

Increasing the size of pins or considering the use of pneumatic air ejection to assist in component release may help to solve the ejection pin marks problem. Flow chart 5.2.17 shows the way to approach the problem of ejector pin marks.

Problems and Troubleshooting – Injection Molding

Flow chart 5.2.17 – Ejector pin marks

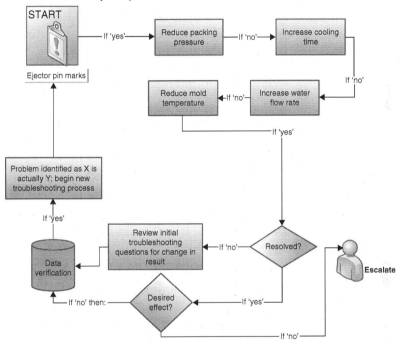

5.2.18   Sink Marks

The first thing to understand about sink marks is a sink or a pull. It can be eliminated or decreased with some simple practices. Study of part weight when the gate freezes and packing of material into the mold is also important. A gate that is too far away or too small may also be a cause of the sink marks. Enlarging the gate or runner may solve the problem of sink marks. Flow chart 5.2.18 shows the way to solve the problem of sink marks.

Residence time is increased with low powder feed rate. Hence, decrease in torque results in restricted packing of material which leads to sink marks or porosity in the end product. Ribs are a major source of sinks found on adjacent walls. Changing rib thickness is a common strategy for reducing or eliminating sinks. An insufficient cushion may cause sink marks on the part. Low holding pressure may lead to sink marks and will increase shrinkage.

Problems and Troubleshooting – Injection Molding
Flow chart 5.2.18 – Sink marks

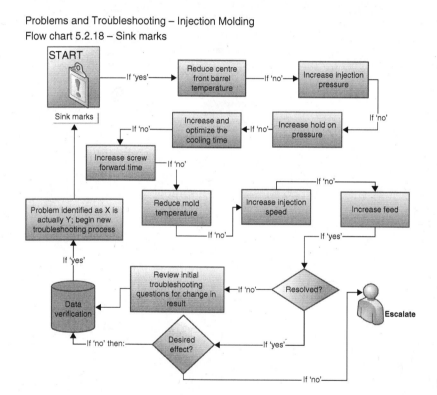

5.2.19 Shrinkage

Shrinkage is a complex parameter and can be influenced by many interrelated factors, such as machinery, mold design, part geometry and direction of flow, processing conditions, and molding system. Shrinkage is critical to control the dimensions of the finished part. Flow chart 5.2.19 shows the way to approach and solve the problem of shrinkage.

The thickness and geometry of the part play an important role in determining the shrinkage. The shrinkage is higher in the flow direction than perpendicular to the flow direction in most of the plastic materials. Influence of processing conditions on shrinkage and shrinkage may be very difficult to change. Processing conditions have an influence on shrinkage but can be varied.

Problems and Troubleshooting – Injection Molding
Flow chart 5.2.19 – Shrinkage

Higher mold temperature will result in shrinkage increase. Melt temperature does not have influence on shrinkage. Higher and long holding pressures will result in less shrinkage.

Shrinkage is due to non-uniform cooling and packing throughout the cavity. It is further affected by mold cooling, constraints of mold geometry, and possible presence of filler.

The residual stresses are produced due to high pressure and cooling, which include warpage and shrinkage. The shrinkage and warpage build up during the solidification of thermoplastic melts [25, 26].

5.2.20   Silver Streak

Too high shear can often cause silver streaking. Controlling and reducing the flow rate through the area where the splay is seen can eliminate this. Sometimes machine capacity may be higher rated than required; in such case a lower capacity machine may help to reduce silver streak. Flow chart 5.2.20 shows the way to solve the problem of silver streak.

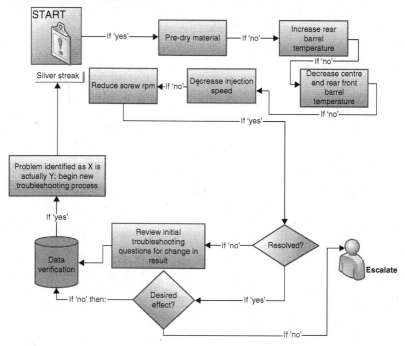

Problems and Troubleshooting – Injection Molding

Flow chart 5.2.20 – Silver streak

5.2.21   Short Shots

Short shots are when enough polymer melt is not flowed into a cavity to fill it adequately. To fill the cavity, the determination

Problems and Troubleshooting – Injection Molding
Flow chart 5.2.21 – Short shots

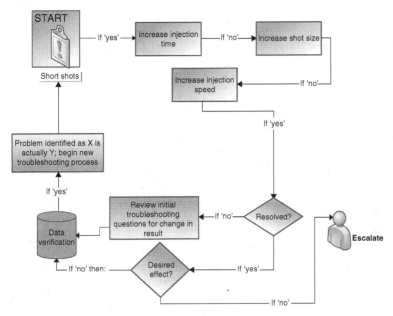

of pressure is required to start pack and hold of the material until gate freeze occurs. During screw rotation, there may be a possibility of letting material leak past the valve. Decompression increase may allow the valve to seat and shut off condition better. But it could cause splay or other cosmetic problems. Slow injection rate sometimes causes the problem of shot size. Increasing the gate and deepening the vents may solve the problem. Flow chart 5.2.21 illustrates the schematic approach to solve the problem of short shots. Short shot problems occur due to high viscosity characteristics of polymer melt during manufacturing.

5.2.22   Brittle Part

Contamination may cause brittleness; raw material checked for contamination and at the same time regrinds level, if any, reduction will help to solve the problem. Too much additive

Problems and Troubleshooting – Injection Molding
Flow chart 5.2.22 – Brittle part

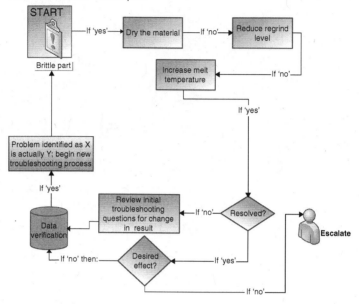

or concentrates can cause the parts to be brittle. Melt temperature is within the specified range. High melt temperature can degrade the material to a point where it will become brittle. Flow chart 5.2.22 shows the problem of brittle part and the approach to solve it.

5.2.23   Poor Plasticizing

Flow chart 5.2.23 shows the solution to the problem of poor plasticizing. Poor plasticizing starts from the problem either of temperature or material melting. Poor plasticizing reduces the mechanical properties of the end products.

5.2.24   Crack During Mold Release

Flow chart 5.2.24 shows the schematic approach to the problem of crack during mold release. This type of crack may happen during mold release.

Problems and Troubleshooting – Injection Molding

Flow chart 5.2.23 – Poor plasticizing

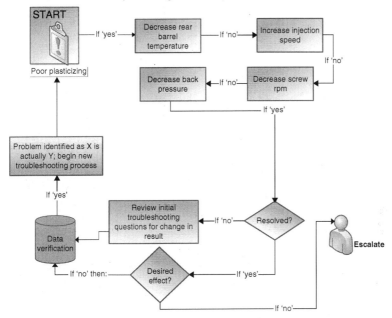

Problems and Troubleshooting – Injection Molding

Flow chart 5.2.24 – Crack during mold release

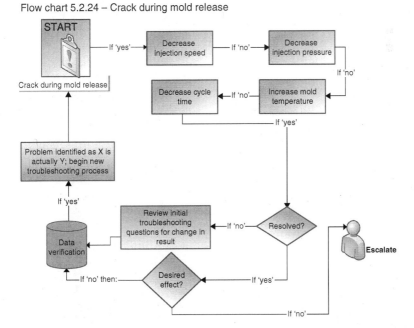

5.2.25 Nozzle Drool

Using nozzle with smaller diameter orifice or decompression may help to reduce the nozzle drool. Flow chart 5.2.25 shows the solution to solve the problem of nozzle drool.

5.2.26 Short Shots No Burn Marks

Inspecting and cleaning the vents may solve the problem. Decision of machine capacity is a must before any part has to be injection molded. These types of short shots will be basically not having enough capacity to fill the cavity in most of the cases. Flow chart 5.2.26 shows the schematic approach to solve the problem of short shots with no burn marks.

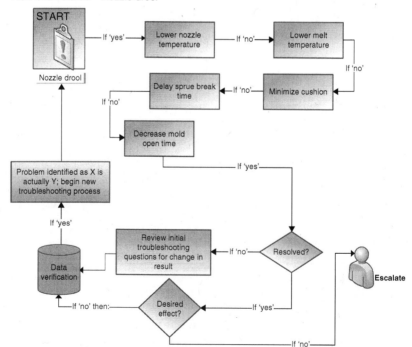

Problems and Troubleshooting – Injection Molding

Flow chart 5.2.25 – Nozzle drool

Problems and Troubleshooting – Injection Molding

Flow chart 5.2.26 – Short shots no burn mark

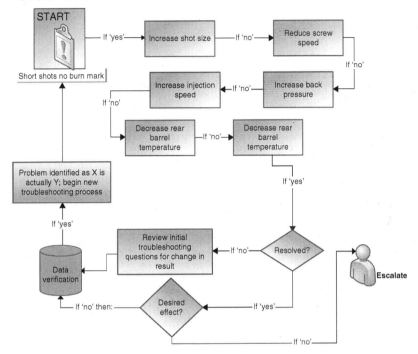

5.2.27   Surface Ripples, and Pit Marks

Purging the machine and reducing the recycle if added may resolve the problem. Using large capacity machine and addition of vents also solves the problem. In any problem, leakage of check ring may increase the problem. Flow chart 5.2.27 shows the way to solve the problem of surface ripples and pit marks.

5.2.28   Pellets Not Melted

Machine plasticizing capacity may be too small for shot and cycle or may not melt sufficient material for the required injection. Worn out screw barrel or not having proper compression ratio in the screw or screw flight depth may cause the melting related problems. Flow chart 5.2.28 illustrates the schematic approach to solve the problem of pellets not melted.

Problems and Troubleshooting – Injection Molding

Flow chart 5.2.27 – Surface ripples and pit marks

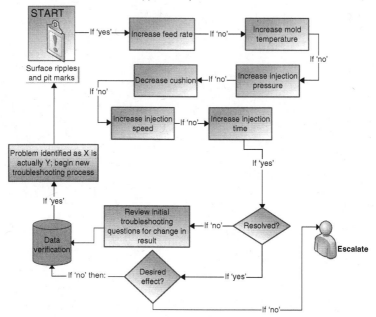

Problems and Troubleshooting – Injection Molding

Flow chart 5.2.28 – Pellets not melted

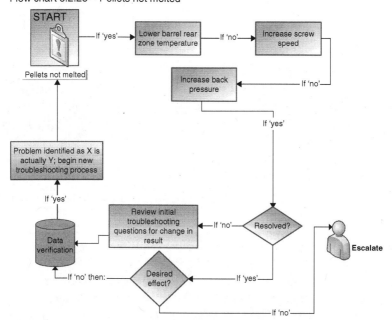

5.2.29   Air Entrapment in the Mold

Air entrapment in the mold will prevent the mold cavity from filling properly. Air pockets have particularly negative influence on the final product. Proper venting has to be facilitated in the mold to avoid the problem of air entrapment. Flow chart 5.2.29 illustrates the schematic approach to solve the problem of air entrapment in the mold. In some cases, two or more material flow fronts are converging on each other, which can sometimes cause a gas trap that is eventually moved from the area and is smeared out across the part. Elimination can be done by changing the material flow in that area.

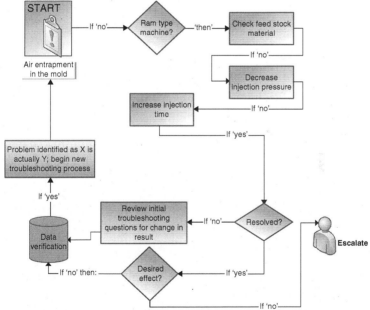

Problems and Troubleshooting – Injection Molding
Flow chart 5.2.29 – Air entrapment in the mold

Trapped air or volatiles can prevent good knitting of converging flow fronts. Core pins, "blind" holes, and special mold features can cause air entrapment. Jetting can also be a cause of non uniform melt flow and weak flow front melding.

Insufficient venting results in compression of air trapped in the cavity. This results in very high temperatures and pressures in the cavity, causing burn marks, material degradation, and short shots.

Sometimes volatiles emitted by the resin during processing decrease weld-line strength. Unless properly vented, the gas can hold the flow fronts apart.

5.2.30   Gate Blush

Gate blush is very common in injection molding and is seen with the use of sub-gates or cashew gates. Flow chart 5.2.30 illustrates the schematic approach to solve the problem of gate blush. Over pack of the part can cause a blush to be seen around the gate area. High flow rate can sometimes result in a gate blush. Lower injection rate and making the gate larger can help to eliminate the problem.

Blush is a cosmetic flaw (dull spots) caused by improper plastic dispersion during injection [27].

Problems and Troubleshooting – Injection Molding

Flow chart 5.2.30 – Gate blush

5.3   Troubleshooting – Extrusion

Melt flow properties are useful in selecting an appropriate extruder screw and die, in setting appropriate processing conditions in troubleshooting extrusion problems, and in allowing prediction of extrusion behavior.

Die exit surface defects include scratches, loss of transparency, and cracks at low flow rates. [28].

Extrusion instabilities

- Frequency of instabilities – low and high frequency, screw instabilities, very slow and random fluctuations
- Functional instabilities – feed problems, melting related problems, melt conveying problems, mixing related instabilities, solving extrusion instabilities.

Air entrapment

- Removing entrapped air
- Avoiding air entrapment.

Die flow problems

- Melt fracture
- Die lip buildup
- Specks and discoloration
- Die lines and weld lines.

5.3.1   "Bridging" at the Throat of the Feed Hopper

Bridging at the feed hopper throat comes from the use of higher percentage of waste material or regrind in blend with virgin material. Excessive ultrafine powder in the material may also cause bridging. The regrind may be poorly dispersed with the virgin material.

Problems and Troubleshooting – Extrusion

Flow chart 5.3.1 – "Bridging" at the throat of the feed hopper

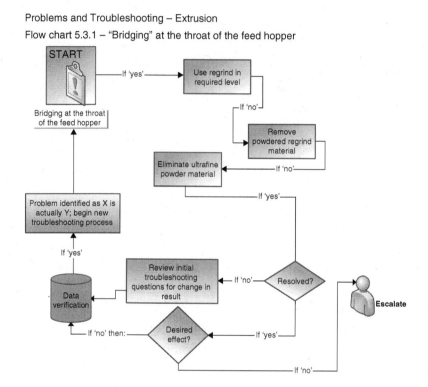

5.3.2   Brittle/Braking/Tearing

Contamination of foreign material or thermoplastics material (not possible to blend) may cause brittleness or braking or tearing of extruded materials. Excessive orientation in the direction of extrusion can cause tearing. In the film processing, excessive cooling on the polish roll may break while extruding. Large agglomeration of additives such as pigment can cause brittleness. Incorrect choice of material or material inhomogeneity might cause melt strength problems which leads to braking of the polymer melt.

Problems and Troubleshooting – Extrusion

Flow chart 5.3.2 – Brittle/braking/tearing

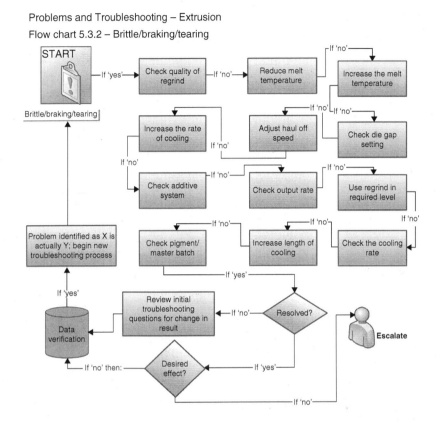

5.3.3   Contamination/Black Specks and Spots

Black specks and spots or contamination result from degraded material or foreign material contamination. Contamination of other thermoplastics also causes the problem. The re-use of regrind is effectively managed to use in a closed distribution system. The regrind must be stored in a better way. Screen pack addition in the machine with suitable mesh size filters the contamination, but it requires frequent cleaning or change of screen. Non-homogenous material causes black specks and spots. Hold ups or dead spots along screws or in the die can cause the black specks. Die should be cleaned and free from any carbonized material.

Problems and Troubleshooting – Extrusion

Flow chart 5.3.3 – Contamination/Black specks and spots

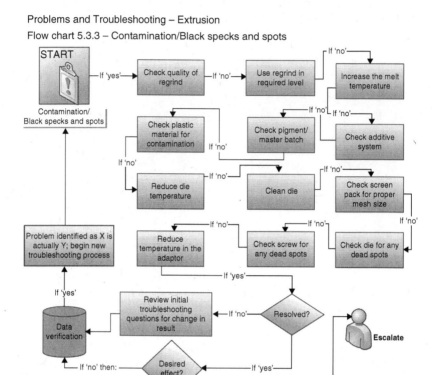

5.3.4   Die Lines

Die lines may be from build-up of degraded carbonized debris on the die lip or in the die head. The die lines may also start from the physically damaged die lips.

5.3.5   Entrapped Gasses/Air Bubbles

Entrapment of gas in the product may come from the volatiles or air. Moisture may also cause entrapment along with the material.

5.3.6   Excessive Die Swell

Material may not be suitable to extrude or it is subjected to high levels of shear which leads to over melting and their associated problems.

Problems and Troubleshooting – Extrusion
Flow chart 5.3.4 – Die lines

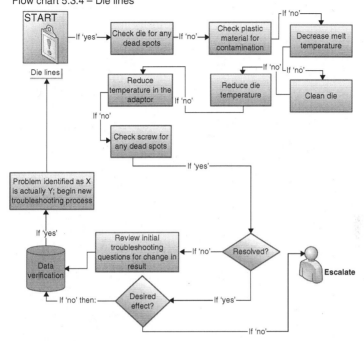

Problems and Troubleshooting – Extrusion
Flow chart 5.3.5 – Entrapped gasses/air bubbles

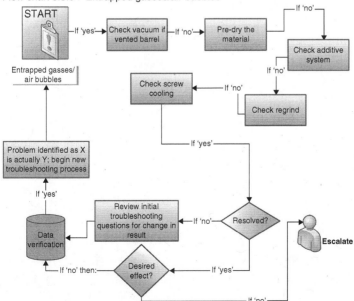

Problems and Troubleshooting – Extrusion

Flow chart 5.3.6 – Excessive die swell

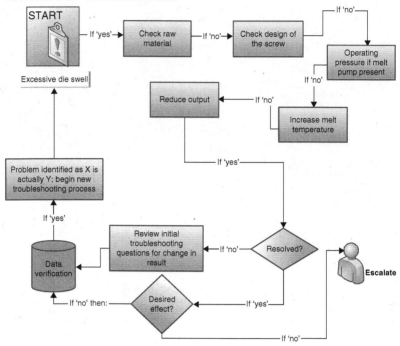

5.3.7   Material Accumulation at Die

Thickness setting for the die gap may be too high. The take-off/haul-off function is not synchronized properly to take off the melt from the die.

5.3.8   Low Gloss

In extrusion, low gloss comes out from the die and polymer melt temperature. Surface finish of the die is quite important and helps to give gloss to the product.

5.3.9   Material Non-Homogeneous

Screw not optimized for the purpose of using the particular thermoplastics leads to material non-homogeneous in nature.

Problems and Troubleshooting – Extrusion
Flow chart 5.3.7 – Material accumulation at the die

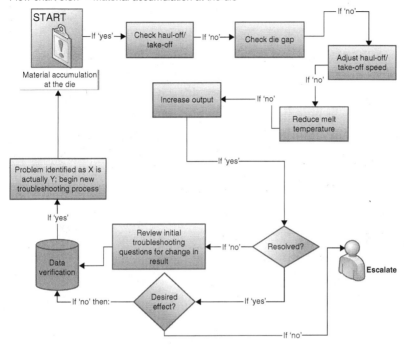

Problems and Troubleshooting – Extrusion

Flow chart 5.3.8 – Low gloss

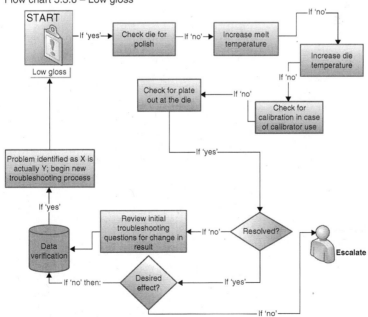

Low melt temperature with high output can cause only shear heating. The shear heat causes the material to not properly melt and includes particles within melt.

Problems and Troubleshooting – Extrusion

Flow chart 5.3.9 – Material non-homogeneous

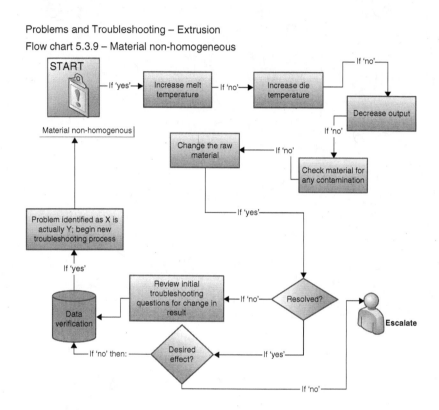

5.3.10 Variable Thickness

Die gap setting has to be correct; otherwise, it results in thickness variation. Material melt temperature increase can cause flow variation in die which is reflected in the variation.

5.3.11 Variable Output/Surging

Surging occurs due to low molecular weight material or regrind addition may be higher in percentage. Increase in melt temperature can also cause surging.

Problems and Troubleshooting – Extrusion
Flow chart 5.3.10 – Thickness variation

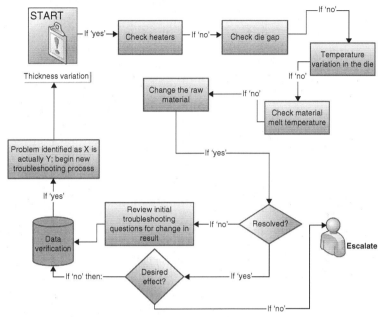

Problems and Troubleshooting – Extrusion
Flow chart 5.3.11 – Variable output/surging

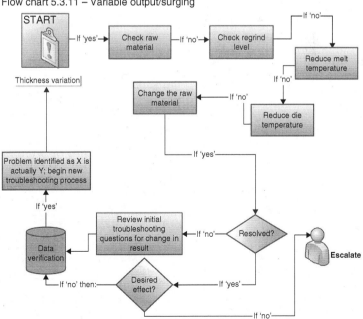

5.3.12   Plate-Out on Die Lip

Plate-out results from the additives such as filler. Fillers will create plate-out due to their non-homogeneous nature with thermoplastics. High melt temperature results in degradation of the material and can cause plate-out on die lip.

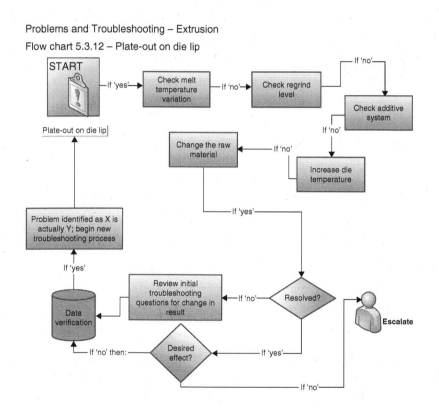

Problems and Troubleshooting – Extrusion

Flow chart 5.3.12 – Plate-out on die lip

5.4   Troubleshooting – Blow Molding

Troubleshooting of blow molding occurs due to machine, material, and die. In extrusion blow molding, the problems occur from the parison and mold adjustment, material properties, and flow rates. In injection stretch blow molding, initially the injection molded product will be initially produced and then stretched to blow in the next stage. In injection blow molding the stretching problem occurs due to the heater, mold, air problem. Surface problem normally occurs in both processes.

5.4.1   Extrusion Blow Molding

5.4.1.1   Curling

Problems and Troubleshooting – Extrusion blow molding

Flow chart 5.4.1.1 – Curling

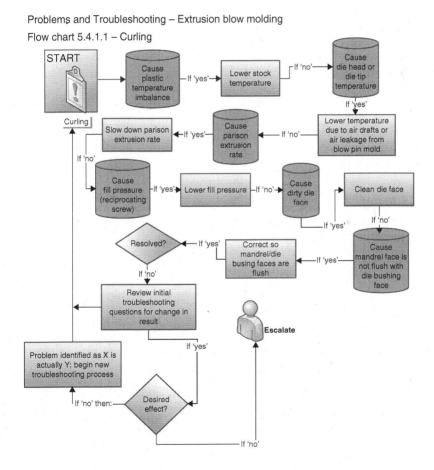

Curling occurs due to non-alignment of parison from the die head. Flow chart 5.4.1.1 shows how to rectify the problem of curling.

5.4.1.2   Hooking

A hooking problem occurs due to parison thickness imbalance. Flow chart 5.4.1.2 indicates the solution to rectify the problem of hooking.

Problems and Troubleshooting – Extrusion blow molding

Flow chart 5.4.1.2 – Hooking

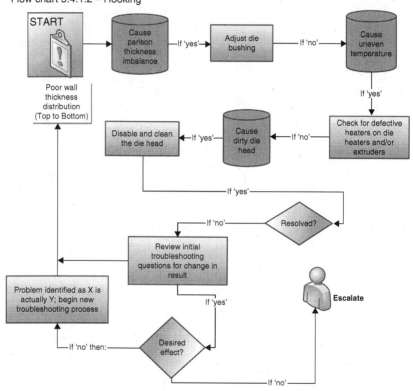

5.4.1.3   Inconsistent Tail Length

Improper material flow may lead to inconsistent tail length. Flow chart 5.4.1.3 indicates the solutions to solve the problem of inconsistent tail length.

5.4.1.4   Blowouts or Pin Holes

The material may contain contamination which causes blowouts or pinholes. Flow chart 5.4.1.4 indicates the solution to the problem of blowouts or pinholes.

Problems and Troubleshooting – Extrusion blow molding

Flow chart 5.4.1.3 – Inconsistent tail length

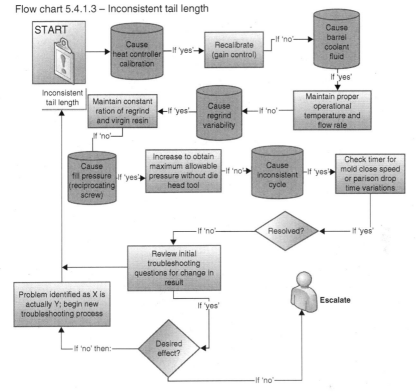

Problems and Troubleshooting – Blow molding

Flow chart 5.4.1.4 – Blowouts or pin holes

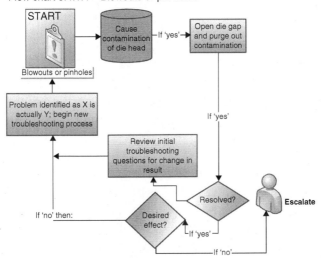

5.4.1.5   Poor Wall Thickness Distribution (Top to Bottom)

Selection of material may not suit and hence the thickness variation occurs from the parison. Flow chart 5.4.1.5 indicates the solution to the wall thickness distribution problem.

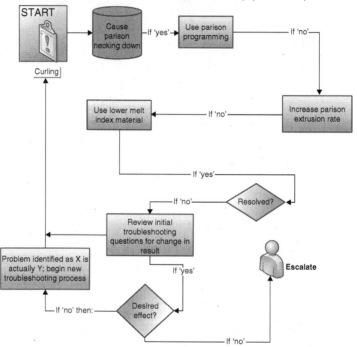

Problems and Troubleshooting – Extrusion blow molding
Flow chart 5.4.1.5 – Poor wall thickness distribution (top to bottom)

5.4.1.6   Asymmetric Part

The air pressure pin may be choked and has to be checked to see that air pressure freely comes from the pin to inflate the parison. Flow chart 5.4.1.6 indicates how to solve the problem of asymmetric part.

5.4.1.7   Poor Weld

Inflated material inside the die cooled very fast causes poor weld problem. Flow chart 5.4.1.7 indicates how to solve the problem of poor weld.

Problems and Troubleshooting – Extrusion blow molding

Flow chart 5.4.1.6 – Asymmetrical part

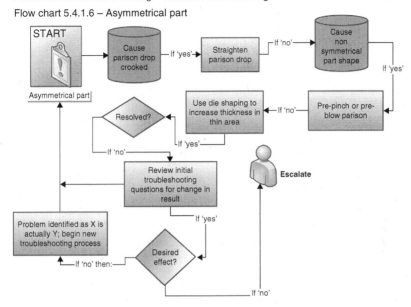

Problems and Troubleshooting – Extrusion blow molding

Flow chart 5.4.1.7 – Poor weld

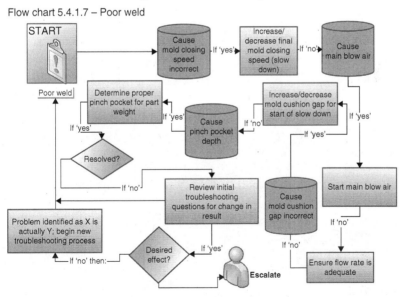

5.4.1.8   Poor Parting Line

The part gets cracked due to poor parting line. Flow chart 5.4.1.8 indicates the solution to the problem of poor parting line.

Problems and Troubleshooting – Extrusion blow molding
Flow chart 5.4.1.8 – Poor parting line

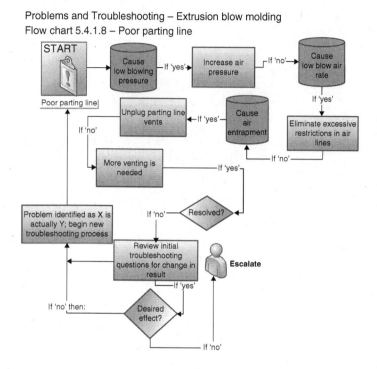

5.4.1.9   Contamination in Parts

Dirty material or improperly stored regrind causes the problem. Flow chart 5.4.1.9 illustrates the solution to solve the contamination in parts.

5.4.1.10   Flashing Tear

The material gets cooled immediately in the mold. Flow chart 5.4.1.10 indicates solution to solve the flashing tear problem.

5.4.1.11   Holes in Pinch-Offs

Damaged pinch-off may cause the problem of holes in a part. Flow chart 5.4.11 indicates the problem of holes in pinch-offs problem.

5.4.1.12   Blowouts

The material may have contamination. Flow chart 5.4.12 indicates solutions to solve the problem of blowouts.

Problems and Troubleshooting – Extrusion blow molding
Flow chart 5.4.1.9 – Contamination in parts

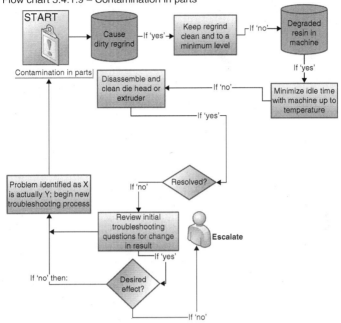

Problems and Troubleshooting – Extrusion blow molding
Flow chart 10 – Flashing tear

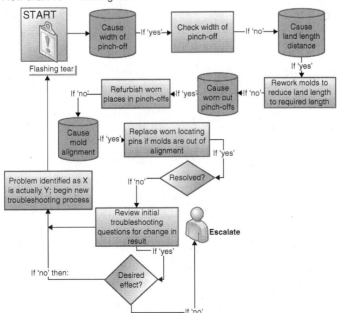

Problems and Troubleshooting – Extrusion blow molding
Flow chart 5.4.1.11 – Holes in pinch-offs

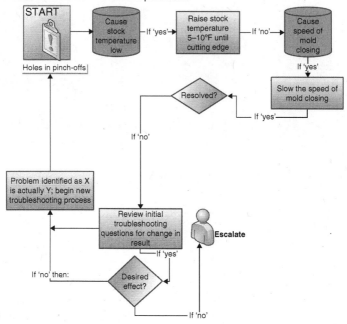

Problems and Troubleshooting – Extrusion blow molding
Flow chart 5.4.1.12 – Blow-outs

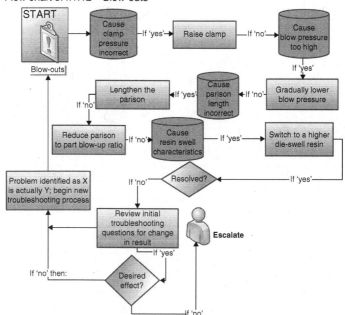

5.4.2   Injection Blow Molding

5.4.2.1   Poor Gates (Fish Eyes, Flash, Tails, etc.)

One of the reasons may be the contamination which causes the poor gates with fish eyes, flash, tails, etc. Flow chart 5.4.2.1 indicates a solution to the problem of poor gates (fish eyes, flash, tails, etc).

Problems and Troubleshooting – Injection blow molding
Flow chart 5.4.2.1 – Poor gates (fish eyes, flash, tails, etc.)

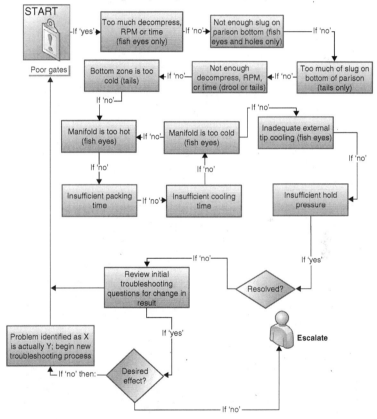

5.4.2.2   Poor Gates

Improper vent and insufficient blow time may cause poor gates. Flow chart 5.4.2.2 indicates the solution to the poor gates problem.

Problems and Troubleshooting – Injection blow molding
Flow chart 5.4.2.2 – Poor gates

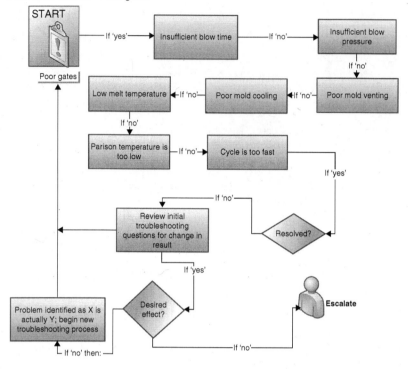

5.4.2.3   Rocker Bottoms

Increase in cycle time and excess temperature may cause rocker bottoms. Flow chart 5.4.2.3 indicates the solution to rocker bottoms problem.

5.4.2.4   Incomplete Thread

Excess mold temperature causes the problem of incomplete thread. Flow chart 5.4.2.4 indicates the solution to solve the problem of incomplete thread.

5.4.2.5   Pig Tails

Excessive material may be sticking, causing the problem of pig tails. Flow chart 5.4.2.5 indicates the solution to the problem of pig tails.

Problems and Troubleshooting – Injection blow molding
Flow chart 5.4.2.3 – Rocker bottoms

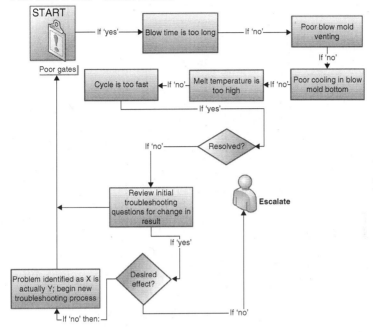

Problems and Troubleshooting – Injection blow molding
Flow chart 5.4.2.4 – Incomplete threads

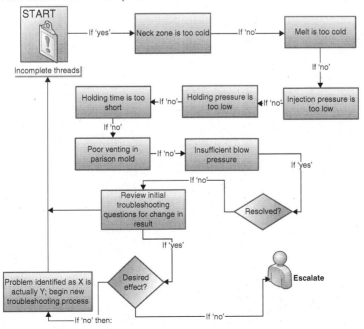

Problems and Troubleshooting – Injection blow molding
Flow chart 5.4.2.5 – Pig tails

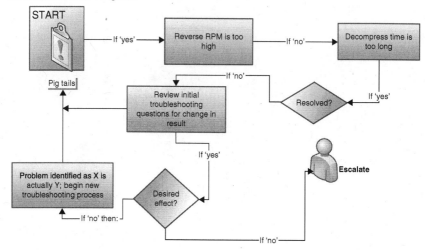

5.4.2.6   Short Shots

Polymer melt flow rate may be lower or material may not feed properly. Flow chart 5.4.2.6 indicates the solution to the problem of short shots.

5.4.2.7   Parison Flashing

Overheating of material is one of the causes to the problem of parison flashing. Flow chart 5.4.2.7 indicates the solution to the problem of parison flashing.

5.4.2.8   Neck Folds – Shoulder Cuts

Mold temperature too low may cause the problem. Flow chart 5.4.2.8 indicates the solution to the problem of neck folds – shoulder cuts.

5.4.2.9   Plastic Sticking to Core Rods

Excessive mold temperature causes the problem. Flow chart 5.4.2.9 indicates the solution to the problem of plastic sticking to core rods.

Problems and Troubleshooting – Injection blow molding
Flow chart 5.4.2.6 – Short shots

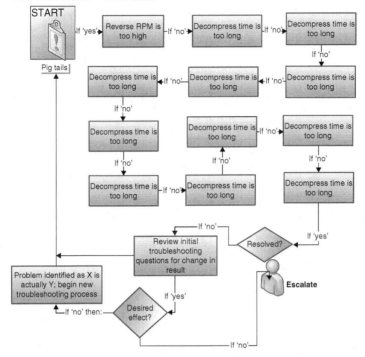

Problems and Troubleshooting – Injection blow molding
Flow chart 5.4.2.7 – Parison flashing

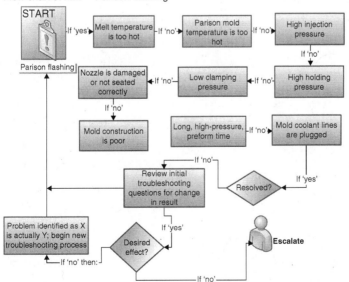

Problems and Troubleshooting – Injection blow molding
Flow chart 5.4.2.8 – Neck folds – Shoulder cuts

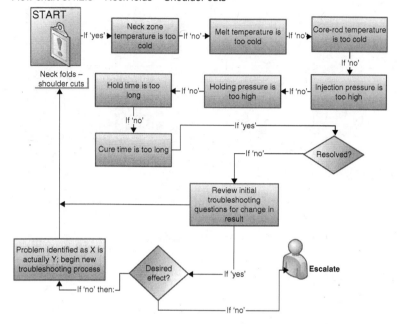

Problems and Troubleshooting – Injection blow molding
Flow chart 5.4.2.9 – Plastic sticking to core rods

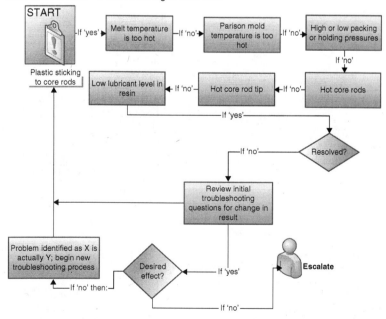

5.5   Troubleshooting – Thermoforming

Product consistency from thermoforming depends on controlling variables of material and process. The important variable is the quality of the sheet feed stock. The extruded sheet of the same material could vary in terms of polymeric contamination, thickness, thermal stresses, and amount of regrind, volatiles, color, gloss, and grain. Variation in mold material and mold temperature also affects the product consistency. Obviously, during actual thermoforming, many times the effects of such variation troubles occur in the form of tearing, wall thinning, shape distortion, fading, pinholes, and grain distortion [29].

5.5.1   Blisters or Bubbles

Presence of moistures or volatiles causes the problem of blisters or bubbles. Flow chart 5.5.1. indicates the way to solve the problem.

5.5.2   Incomplete Forming or Poor Detail

Temperature may not be sufficient to heat the sheet material. Flow chart 5.5.2. indicates solutions to the problem of incomplete forming or poor detail.

5.5.3   Sheet Scorched

Use suitable heaters to heat the sheet. Flow chart 5.5.3. indicates solutions to solve the problem of sheet scorched.

5.5.4   Changing in Color Intensity or Blushing

Low heating of the sheet material or cycle time too long causes the problem. Flow chart 5.5.4 indicates solutions to solve the problem of changing in color intensity or blushing.

Problems and troubleshooting – Thermoforming
Flow chart 5.5.1 – Blisters or bubbles

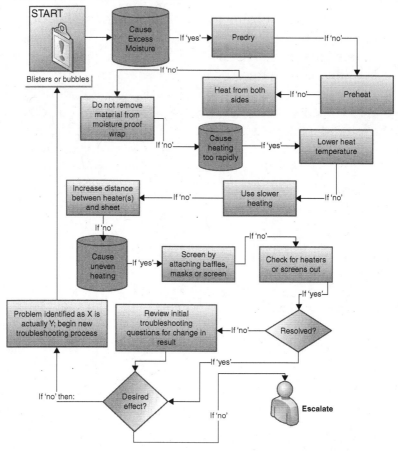

5.5.5   Whitening of Sheet

Too high stress is developed on the sheet material. Reducing the stress may solve the problem. However flow chart 5.5.5 helps to resolve the problem of whitening of sheet.

5.5.6   Webbing, Bridging or Wrinkling

Reduce the distance between the sheet material and heating area. Flow chart 5.5.6 indicates the solution to resolve the problem of webbing.

Problems and troubleshooting – Thermoforming
Flow chart 5.5.2 – Incomplete forming or poor detail

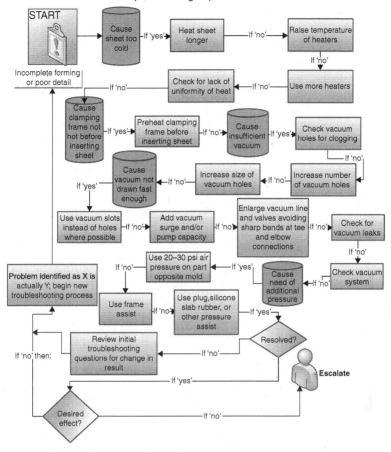

Problems and troubleshooting – Thermoforming
Flow chart 5.5.3 – Sheet scorched

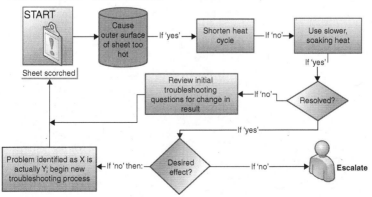

Problems and troubleshooting – Thermoforming
Flow chart 5.5.4 – Changing in color intensity or Blushing

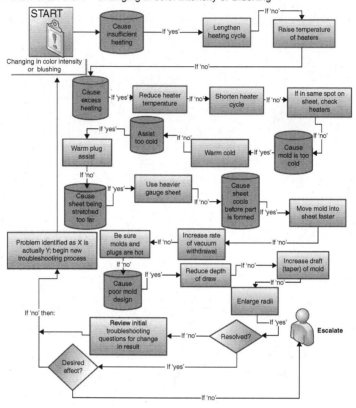

Problems and troubleshooting – Thermoforming
Flow chart 5.5.5 – Whitening of sheet

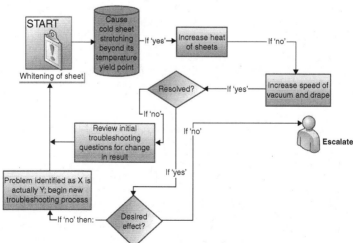

Problems and troubleshooting – Thermoforming

Flow chart 5.5.6 – Webbing, bridging or wrinkling

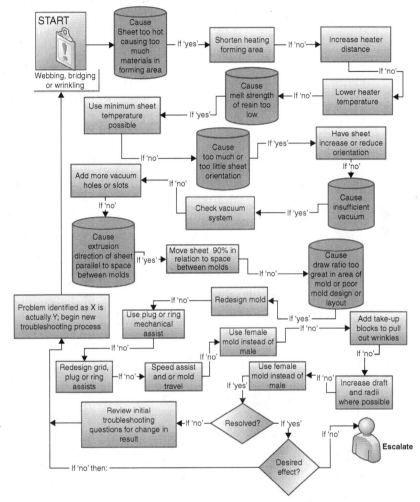

5.5.7   Nipples on Mold Side of Formed Sheet

Too high temperature causes the problem of nipples on mold side of a formed sheet. Flow chart 5.5.7 indicates the solution to solve the problem.

5.5.8   Too Much Sag

A sheet may be made from low molecular material or too high temperature causes the problem of too much sag. Flow chart 5.5.8 indicates how to solve the problem of too much sag during processing.

5.5.9 Sag Variation between Sheet Blanks

Check the sheet material for melt strength. Flow chart 5.5.9 helps to resolve the problem of sag variation between sheet blanks.

5.5.10 Chill Marks or Mark-off Lines on Part

Too high temperature may cause the problem of chill marks. Flow chart 5.5.10 indicates the solution to solve the problem.

5.5.11 Bad Surface Marking

Hot air pockets may cause the problem of bad surface marking. Flow chart 5.5.11 helps to resolve the problem.

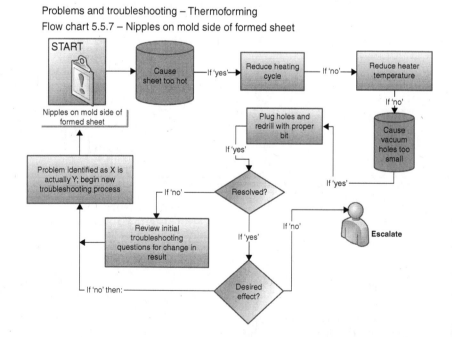

Problems and troubleshooting – Thermoforming
Flow chart 5.5.7 – Nipples on mold side of formed sheet

Problems and troubleshooting – Thermoforming
Flow chart 5.5.8 – Too much sag

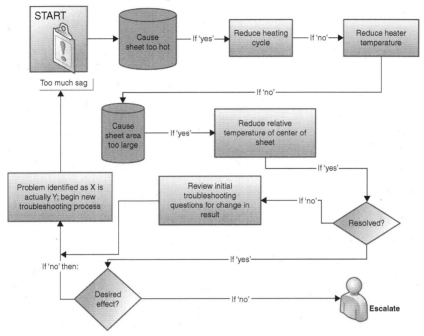

Problems and troubleshooting – Thermoforming
Flow chart 5.5.9 – Sag variation between sheet blanks

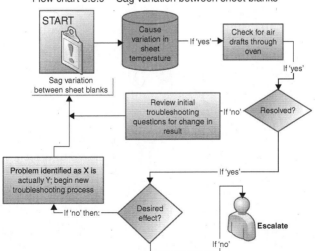

Problems and Troubleshooting – Thermoforming
Flow chart 5.5.10 – Chill marks or mark-off lines on part

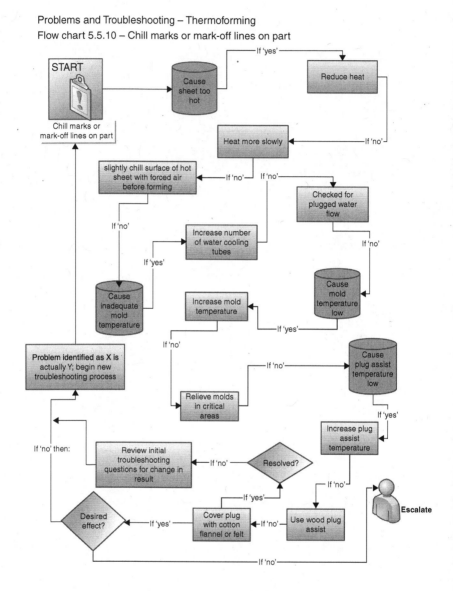

5.5.12   Shiny Streaks on Part

Release agent may be the cause for shiny streaks on a part.
Flow chart 5.5.12 helps to resolve the problem.

Problems and troubleshooting – Thermoforming

Flow chart 5.5.11 – Bad surface markings

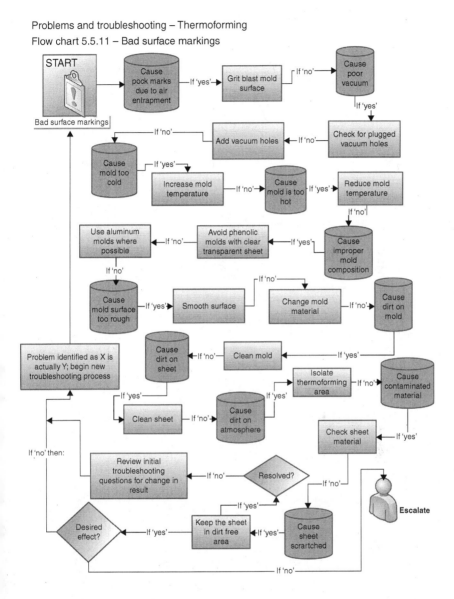

5.5.13   Excessive Post Shrinkage or Distortion of Part Removing from the Mold

Too high temperature or too low cycle time may cause the problem of excessive post shrinkage. Flow chart 5.5.13 helps to resolve the problem.

Problems and troubleshooting – Thermoforming

Flow chart 5.5.12 – Shiny streaks on part

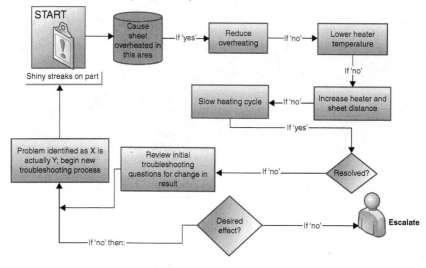

Problems and troubleshooting – Thermoforming

Flow chart 5.5.13 – Excessive post shrinkage or distortion of part removing from mold

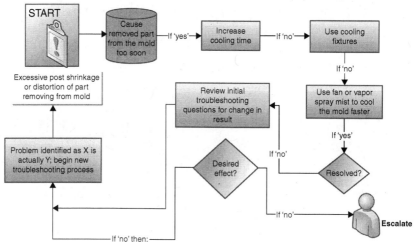

5.5.14 Part Warpage

Too high temperature causes the problem of part warpage. Flow chart 5.5.14 helps to resolve the problem.

Problems and troubleshooting – Thermoforming
Flow chart 5.5.14 – Part warpage

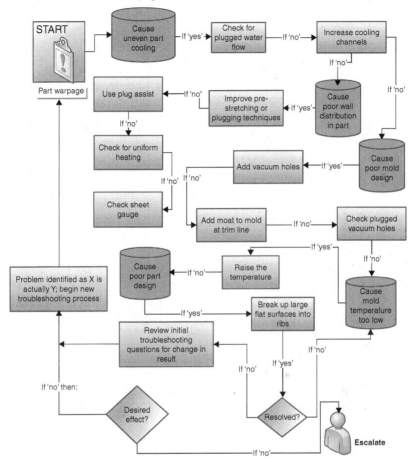

5.5.15   Poor Wall Thickness or Excessive Thinning in Some Areas

Improper temperature causes the problem of poor wall thickness. Flow chart 5.5.15 helps to resolve the problem.

5.5.16   Non-Uniform Pre-stretch Bubble

The heater may be the problem causing a non-uniform pre-stretch bubble problem. Flow chart 5.5.16 helps to resolve the problem.

Problems and Troubleshooting – Thermoforming

Flow chart 5.5.15 – Poor wall thickness or excessive thinning in some areas

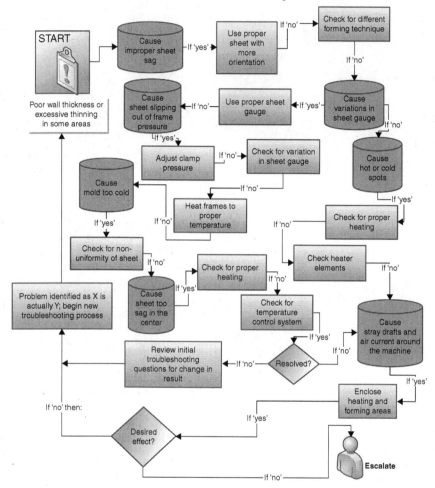

5.5.17 Shrink Marks on Part, Especially in Corner Areas (Inside Radius of Molds)

Vacuum may be chocked or insufficient capacity vacuum may cause the problem. Flow chart 5.5.17 helps to resolve the problem of shrink marks.

Problems and Troubleshooting – Thermoforming
Flow chart 5.5.16 – Non-uniform pre-stretch bubble

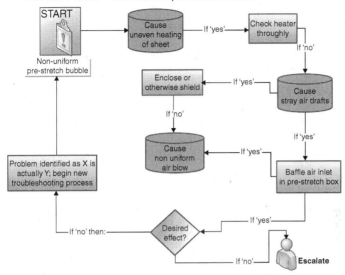

Problems and Troubleshooting – Thermoforming
Flow chart 5.5.17 – Shrink marks on part, especially in
corner areas (inside radius of molds)

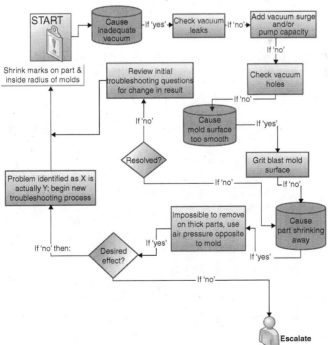

5.5.18   Too Thin Corners in Deep Draws

Thin sheet material may be one of the reasons. Flow chart 5.5.18 helps to resolve the problem of too thin corners in deep draws.

Problems and troubleshooting – Thermoforming

Flow chart 5.5.18 – Too thin corners in deep draws

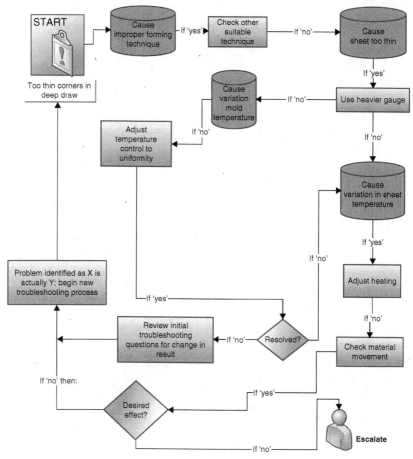

5.5.19   Part Sticking to Mold

Mold temperature too high may cause the problem. Flow chart 5.5.19 helps to resolve the problem of part sticking to mold.

Problems and troubleshooting – Thermoforming

Flow chart 5.5.19 – Part sticking to mold

5.5.20   Sheet Sticking to Plug Assist

Too high temperature can cause the problem of sheet sticking to plug assist. Flow chart 5.5.20 helps to resolve the problem.

5.5.21   Tearing of Part When Forming

Check sheet material characteristics for proper melt strength. Flow chart 5.5.21 helps to resolve the problem of tearing of part when forming.

Problems and troubleshooting – Thermoforming

Flow chart 5.5.20 – Sheet sticking to plug assist

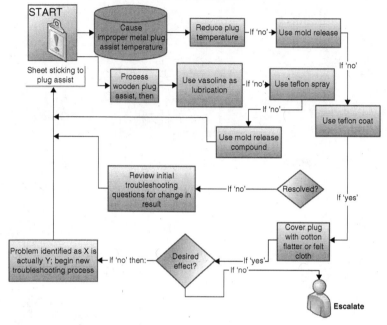

Problems and troubleshooting – Thermoforming

Flow chart 5.5.21 – Tearing of part when forming

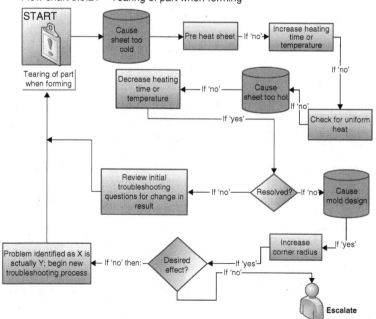

5.5.22   Cracking in Corners During Service

Too low cycle time may cause the problem of cracking in corners during service. Flow chart 5.5.22 helps to resolve the problem.

Problems and Troubleshooting – Thermoforming

Flow chart 5.5.22 – Cracking in corners during service

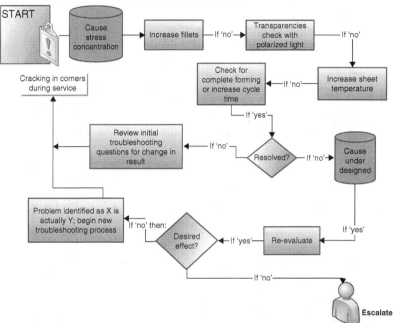

5.6   Troubleshooting – Rotational Molding

The rotational molding processing method for plastics is steadily expanding its product range and market areas. However, a drawback of the process has always been the surface pin-holes and internal bubbles that occur in the products because of the inherent nature of the process [30].

5.6.1   Bubbles on Outer Wall

Check moisture in the master batch. Flow chart 5.6.1 helps to resolve the problem of bubbles.

Problems andTroubleshooting – Rotational molding
Flow chart 5.6.1 – Bubbles on outer wall

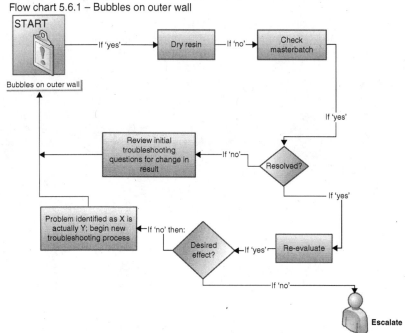

5.6.2   Discolored Part

Temperature too high causes the problem of discoloration. Flow chart 5.6.2 helps to resolve the problem of discoloration.

5.6.3   Flash Excessive

Too high vacuum causes the excessive flash. Flow chart 5.6.3 helps to resolve the problem of excessive flash.

5.6.4   Long Oven Cycle

Heater may not be working properly. Flow chart 5.6.4 indicates how to resolve the problem of long oven cycle.

5.6.5   Low Density Less Than Estimated

Gas may be released while processing or degradation occurred inside the mold. Flow chart 5.6.5 indicates the solutions to solve the problem.

Problems and Troubleshooting – Rotational molding

Flow chart 5.6.2 – Discolored part

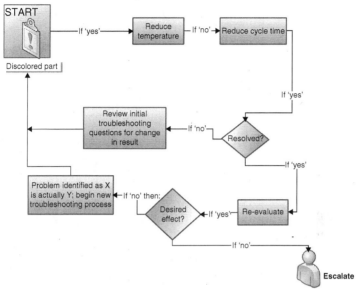

Problems and Troubleshooting – Rotational molding

Flow chart 5.6.3 – Flash excessive

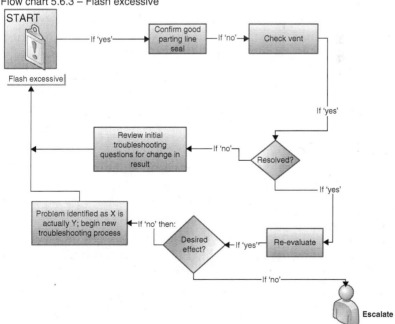

Problems and Troubleshooting – Rotational molding

Flow chart 5.6.4 – Long oven cycle

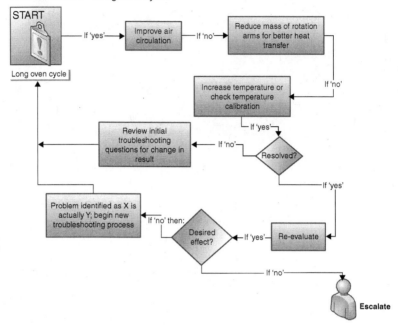

Problems and Troubleshooting – Rotational molding

Flow chart 5.6.5 – Low density less than estimated

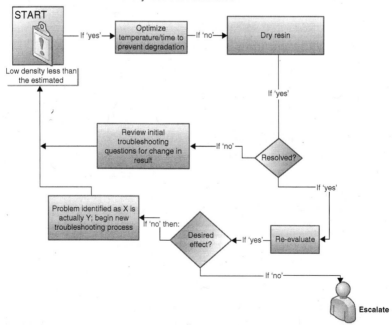

5.6.6 Poor Mold Filling

Low temperature operation causes the problem of poor mold filling. Flow chart 5.6.6 helps to resolve the problem.

Problems and Troubleshooting – Rotational molding

Flow chart 5.6.6 – Poor mold filling

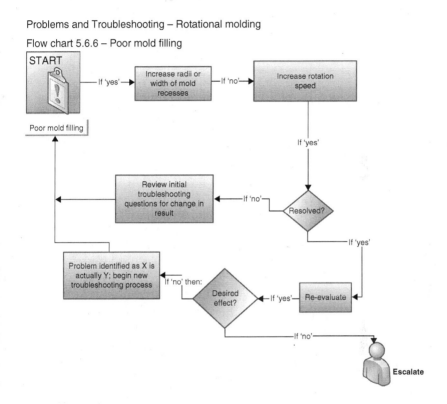

5.6.7 Poor Properties

Too high temperature leads to degradation of the sheet material. Flow chart 5.6.7 helps to resolve the problems of poor properties.

5.6.8 Rough Inner Surface

Low temperature creates rough inner surface. Flow chart 5.6.8 helps to resolve the problem of rough inner surface.

Problems and Troubleshooting – Rotational molding
Flow chart 5.6.7 – Poor properties

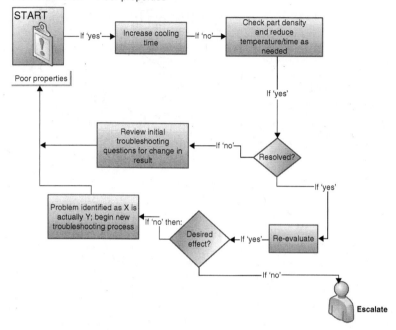

Problems and Troubleshooting – Rotational molding
Flow chart 5.6.8 – Rough inner surface

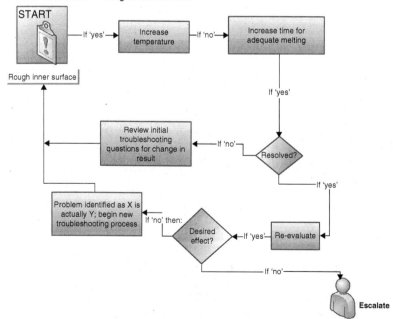

5.6.9   Surface Pitting

Surface pitting occurs due to dirt present in the forming unit. Flow chart 5.6.9 indicates the solution to solve the problem of surface pitting.

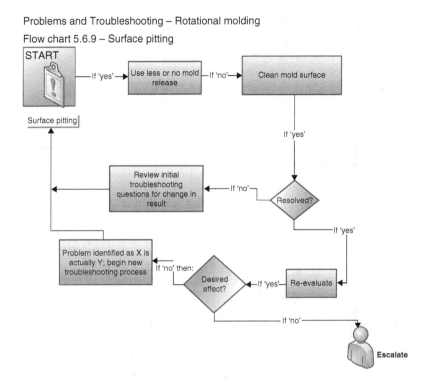

Problems and Troubleshooting – Rotational molding

Flow chart 5.6.9 – Surface pitting

5.6.10   Uneven Wall Thickness

There may be a possibility of frictional forces acting during process. Flow chart 5.6.10 indicates the solution to solve the problem.

5.6.11   Warpage

Insufficient cooling or increase of the cycle time helps to solve the problem. However, flow chart 5.6.11 indicates a solution to the problem of warpage.

Problems and Troubleshooting – Rotational molding

Flow chart 5.6.10 – Uneven wall thickness

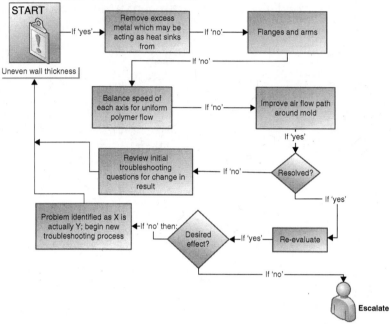

Problems and Troubleshooting – Rotational molding

Flow chart 5.6.11 – Warpage

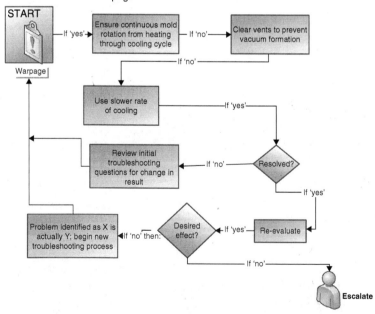

5.7   Fundamentals

- Troubleshooting can reduce long down times, improve product quality and increase efficiency of processing
- Historical data and collected data will help to solve the problem
- The problems start from raw material to end products and process control to processing equipment
- Troubleshooting and problem solving techniques require good understanding of the process
- Efficient troubleshooting requires logical step-by-step approaches to various processing problems
- Troubleshooting requires analysis, approach and attack the problems by appropriate solutions
- Troubleshooting is a guideline to establish the data collection and information related to problem and solution.

References

1. Liu, S.J., Yang, C.Y., *Journal of Reinforced Plastics and Composites* (2004), 23(13), 1383–1396.
2. Richardson, S.M., Pearson, H.J., Pearson, J.R.A., Simulation of injection molding, *Int. J. Plastic Rubber Process.* 5 (1980) 55–65.
3. Hieber, C.A., Shen, S.F., A finite element/finite difference simulation of the injection-molding filling process, *J. Non-Newton. Fluid Mech.* 12 (1980) 1–32.
4. Huilier, D., Leenfant, C., Terrisse, J., Modelling the package state in injection molding of thermo plastics, *Polym. Eng. Sci.* 28 (1988) 1637–1643.
5. Krug, S., et al./*Journal of the European Ceramic Society* 22 (2002) 173–18.
6. Chao-Chyun An, Ren-Haw Chen, *Journal of materials processing technology* 201 (2008) 706–709.
7. Dennison, M.T., Flow instability in polymer melts: a review, *Plastics Polymers*, 35 (1967) 803–808.
8. Tordella, J.P., Unstable flow of molten polymers, in F.R. Eirich (Ed.), *Rheology, Vol. V*, Academic Press, New York, 1969.
9. White, J.L., Critique on flow patterns in polymer fluids at the entrance of a die and instabilities leading to extrudate distortion, *Appl. Polym. Symp.*, 20 (1973) 155–174

10. Petrie, C.J.S. and Denn, M.M., Instabilities in polymer processing, *AIChE J.*, 22 (1976) 209–236.
11. Denn, M.M., Issues in viscolastic fluid mechanics, *Annu. Rev. Fluid Mech.*, 22 (1990) 13–34.
12. Larson, R.G., Instabilities in viscoelastic flows, *Rheol. Acta*, 31 (1992) 213–263.
13. Piau, J.M., Kissi, N. E., Toussaint, F., and Mezghani, A., Distortions of polymer melts extrudates and their elimination using slippery surfaces, *Rheol. Acta*, 34 (1995) 40–57.
14. Pospíil, J., Habicher, W.D., Zweifel, H., Pila, J., Neprek, S., Piringer, G.O., *Addcon World 2001*. 7th International Plastics Additives and Modifiers Conference. Berlin, 2001. Book of Papers, paper 6.
15. Vulic, J., Vitarelli, G., Zenner, J.M., *Macromol Symp* 2001;176:1.
16. Pospíšil, J., Horák, Z., Kruliš, Z., Nešpůrek, S., *Macromol Symp* 1998;135:247.
17. Zhou, H., Xi, G., and Liu, F., Residual Stress Simulation of Injection Molding, *JMEPEG (2008)* 17:422–427.
18. Hill, D., Further studies of the injection moulding process, *Appl. Math. Modelling* 1996, Vol. 20, p. 719 – 730
19. Selden, R., *Polym Engg Sci* 1997;37:205
20. Nadkarni, V.M., Ayodhya, S.R., *Polym Engg Sci* 1993;33:358.
21. Kim, S., Suh, N.P., *Polym Engg Sci* 1986;26:1200.
22. Mielewski, D.F, Bauer, D.R., Schmitz, P.J., Oene, H.V., *Polym Engg Sci* 1998;38:2020.
23. Tomari, K., Tonogal, S., Harada, T., Hamada, H., Lee, K., Morii, T., Meakawa, Z., *Polym Engg Sci* 1990;30:931.
24. Fisa, B., Rahmani, M., *Polym Engg Sci* 1991;31:1330.
25. Bushko, W.C., and Stokes, V.K., Solidification of Thermoviscoelastic Melts. Part I. Formulation of Model Problem, *Polym. Eng. Sci.*, 1995, 35(4), p. 351–364.
26. Bushko, W.C., and Stokes, V.K., Solidification of Thermoviscoelastic Melts. Part II: Effects of Processing Conditions on Shrinkage and Residual Stresses, *Polym. Eng. Sci.*, 1995, 35(4), p. 365–383.
27. Wood, S.L., and Ullman, D.G., The functions of plastic injection moulding features, *Design Studies* 17 (1996) 201–213.
28. El kissi, N., Piau, J.M., Stability phenomena during polymer melt extrusion, in: Piau, J.-M., Agassant, J.-F., (Eds.), *Rheology for Polymer Melts Processing*, Elsevier, Amsterdam, 1996.
29. Dharia, A., *Annual Technical Conference – Society of Plastics Engineers* (2006), 64th 2605–2609.
30. Spence, A.G., and Crawford, R.J., The Effect of Processing Variables on the Formation and Removal of Bubbles in Rotationally Molded Products, *Polymer Engineering and science*, Mid-Apr. 19, Vol. 36, NO. 7.

6

Future Trends

Today, the growth of plastics industry has more than kept pace with the general progress in processing technology. A desire for weight saving and fuel economy is a driving factor, along with other features and characteristics such as cost-effectiveness, design freedom, and corrosion resistance. To the customer and industry, of course, the priority is product quality and industry success is based on a higher production rate with trouble-free run.

6.1 Productivity

Productivity is the primary assignment in the processing industry. Processing industries have performed the task of improving production with process development. Development can be applied by implementing wastage reduction during production. To achieve productivity, application of new technology or adoption of mass production may not always be possible.

Of key importance in any step towards improvement in operation and efficiency is the move from a reactive approach to a proactive approach [1].

6.1.1   Reactive Approach

The traditional reactive approach is always focused on past events, and is about detecting and correcting problems that already exist. It does not require any effort to understand. As an inspection based philosophy, the quality is inspected into products at the expense of high rework or rejects.

6.1.2   Proactive Approach

A proactive approach places importance on measurement, analysis, prediction, and prevention. This approach is needed from the beginning in order to prevent defects and errors appearing later in the product's life cycle. In short, quality must be designed into products and processes rather than inspected into them [2].

In order to increase productivity, it is necessary to

- Minimize the defects
- Optimize the production process
- Investigate the problem
- Improve process
- Improve quality.

6.2   Automotive Applications

Plastics have engineering properties with many versatile processing technologies and present modern product design which is not available in other materials. The use of plastic in automotive components is increased due to computer aided engineering and optimization of tools to be used to design plastic components. However, the use of such tools, particularly optimization methods, to improve the manufacturing process has received much less attention. The cost and complexity of manufacturing automotive plastic products demands the development of optimization methods [3–4].

6.3 Medical Applications

The medical sector is of interest for plastics technologists, because it requires more and more material engineering and properties. The plastics technology has to account for complex biological systems and processes that are far from being understood because of their entire control by nature. Today and in the future, the applications of plastics are essential in surgery, for prosthetic systems, and pharmacology, for drug formulation and controlled drug delivery. In medical applications, plastics are widely used and the desired advances are primarily related to improving their bio-stability and performances. Plastics are remarkable in terms of actual clinical applications due to their hydrophobic nature [5].

6.4 Environmental Issues

The past decade has seen increased awareness of the environmental issues and general support for exploration and implementation of methods and practices to make products and processes more environmentally benign. Environmental issues relate to the use of biopolymers due to the possibility of tailoring the properties within a wide range. Bio-polymers are used in medical applications, where stability, mechanical and thermo-physical properties are concerned. Bio-polymers undoubtedly have a potential for many different applications.

In addition to this, bio-polymers can be easily processed into various products due to their processing properties which are similar to PP or PET. However, bio-polymers easily degrade during thermal processing, which may cause unpredicted performance of the material [6–7].

The potential use of bio-polymers in bulk applications sets out a future need of information from various points of view. The demands on biodegradable materials are, in general, more complicated than that for stable polymers traditionally used in bulk applications [8].

6.5   Fundamentals

1. Productivity is the major requirement in the processing industry. A proactive approach prevents defects and errors
2. Environmental issues relate to the use of bio-polymers. Bio-polymers during processing can degrade and give undesirable performance of the products.

References

1. Kolarik, W.J., and Pan, J.N., (1991), Proactive quality: concept, strategy and tools, Proceedings of the International Industrial Engineering Conference, Norcross, Ga, USA, pp. 411–20.
2. Ross, P.J., (1988), *Taguchi Techniques for Quality Engineering*, McGraw Hill, New York, NY.
3. Tucker, C.L., (eds). *Fundamentals of Computer Modeling for Polymer Processing*. Hanser Publishers: Munich, Germany, 1989.
4. Smith, D.E., *Int. J. Numer. Meth. Engng* 2003;57:1381–1411.
5. Vert, M., *Prog. Polym. Sci.* 32 (2007) 755–761.
6. Kopinke, F.D., Mackenzie, K., Mechanistic aspects of the thermal degradation of poly(lactic acid) and poly(b-hydroxybutyric acid). *J Anal Appl Pyrolysis* 1997;40:43–53.
7. Fan, Y.J., Nishida, H., Shirai, Y., Tokiwa, Y., Endo, T., Thermal degradation behaviour of poly(lactic acid) stereocomplex. *Polym Degrad Stab* 2004;86(2):197–208.
8. Södergärd, A., and Stolt, M., *Prog. Polym. Sci.* 27 (2002), 1123–1163.

Subject Index

221

Also of Interest

Check out these published and forthcoming related titles from Scrivener Publishing

Introduction to Industrial Polyethylene: Properties, Catalysts, Processes by Dennis P. Malpass.

Published 2010. ISBN 978-0-470-62598-9.

Demystifies the largest volume manmade synthetic polymer by distilling the fundamentals of what polyethylene is, how it's made and processed, and what happens to it after its useful life is over.

A Concise Introduction to Additives for Thermoplastic Polymers by Johannes Karl Fink.

Published 2010. ISBN 978-0-470-60955-2.

Written in an accessible and practical style, the book focuses on additives for thermoplastic polymers and describes 21 of the most important and commonly used additives from Plasticizers and Fillers to Optical Brighteners and Anti-Microbial additives. It also includes chapters on safety and hazards, and prediction of service time models.

Handbook of Engineering and Specialty Thermoplastics

Volume One: Polyolefins and Styrenics by Johannes Karl Fink

Published 2010. ISBN 978-0-470-62483-5

Volume Two: Polyethers and Polyesters edited by Sabu Thomas and Visakh P.M. Forthcoming April 2011

Volume Three: Nylons edited by Sabu Thomas and Visakh P.M. Forthcoming April 2011

Volume Four: Water Soluble Polymers edited by Johannes Karl Fink. Forthcoming August 2011.

Polymer Nanoutube Nanocomposites edited by Vikas Mittal. Published 2010. ISBN 978-0-470- 62592-7

Provides a one-stop source for the information on synthesis, properties, and potential applications of nanotube reinforced polymer nanocomposites.

A Guide to Safe Material and Chemical Handling by Nicholas P. Cheremisinoff and Anton Davletshin.

Published 2010. ISBN 978-0-470-62582-8

The volume provides an assembly of useful engineering and properties data on materials of selection for process equipment, and the chemical properties, including toxicity of industrial solvents and chemicals.